高温与吡虫啉交互胁迫对麦长管蚜生活史性状的影响

曹俊宇 / 著

东北林业大学出版社
Northeast Forestry University Press
·哈尔滨·

版权专有　侵权必究
举报电话：0451-82113295

图书在版编目（CIP）数据

高温与吡虫啉交互胁迫对麦长管蚜生活史性状的影响 / 曹俊宇著 . — 哈尔滨：东北林业大学出版社 , 2021.8
 ISBN 978-7-5674-2564-4

Ⅰ.①高… Ⅱ.①曹… Ⅲ.①麦长管蚜－防治 Ⅳ.① S435.122

中国版本图书馆 CIP 数据核字（2021）第 175953 号

责任编辑：许　然
封面设计：马静静
出版发行：东北林业大学出版社
　　　　　（哈尔滨市香坊区哈平六道街 6 号　邮编：150040）
印　　装：三河市德贤弘印务有限公司
规　　格：155 mm × 235 mm　16 开
印　　张：10.25
字　　数：146 千字
版　　次：2022 年 3 月第 1 版
印　　次：2022 年 3 月第 1 次印刷
定　　价：168.00 元

如发现印装质量问题，请与出版社联系调换。（电话：0451-82113296　82191620）

前 言

随着气候变暖的加剧,高温事件的发生会更加频繁、持久,且日最高温的增长趋势也是十分明显的。在自然环境中,生物除了会受到高温胁迫外,农药大量使用所造成的残留同样会对生物的生存带来挑战,因此,关注高温和农药两种常见胁迫对生物的交互影响也是非常现实、必要的。基于此,本书在充分考虑高温类型、胁迫次序以及热历史背景经历的基础上,以小麦蚜虫优势种麦长管蚜为研究对象,以刺吸式口器害虫常用烟碱类化学农药吡虫啉为试验药剂,系统研究了高温事件和化学农药交互胁迫对麦长管蚜核心生命参数、种群适合度和跨代生活史表现的影响,以及热历史背景经历对该虫响应化学农药胁迫的影响。研究结果可为理解气候变暖和化学农药交互胁迫对昆虫影响和生物多样性保护提供数据证据,也可为未来这两种胁迫背景下的麦长管蚜综合防控提供理论依据。所得主要结论如下:

(1)亚致死剂量的高温强度和行为反应时间组合会产生不同的生态学后果。麦长管蚜经历短期 1 d 的三种高温强度、行为反应时间组合 34℃/180 min、36℃/30 min、38℃/10 min 单作处理时,行为反应时间较长的温和和中等高温组合(34℃/180 min 和 36℃/30 min)对当代生活史性状以及子代发育、存活产生的负面影响较大。但随着持续天数的增加,较高高温强度、行为反应时间组合 38℃/10 min 产生的负面影响逐渐增加。

（2）高温强度、行为反应时间组合与吡虫啉同时作用产生复杂的互作效应。温和和中等高温、行为反应时间组合与吡虫啉作用产生较显著的加性或协同负面互作效果，如显著增加当代即时死亡率和适合度性状，显著抑制子代若蚜发育历期和存活；但高强度高温、行为反应时间组合和农药互作，对当代以及子代的生活史性状多不产生显著负面影响，对某些性状甚至会产生有益的拮抗作用，如加速子代发育、提高子代存活率以及刺激子代成蚜繁殖等。

（3）高温、吡虫啉胁迫次序对当代没有显著影响，但产生了显著的跨代效应。麦长管蚜经高温、吡虫啉先后不同胁迫次序处理后，会产生不同的互作效应，对母代本身后续的生活史影响不大。但会产生不同的跨代效应，具体而言，先农药后高温胁迫次序对子代存活产生的负面影响更加强烈，但对子代其他生活史性状产生的影响较小。而先高温后农药胁迫次序尽管对子代存活产生的负面影响较小，但对子代后续的适合度性状影响较大。

（4）高温、吡虫啉双重胁迫比单一高温胁迫对麦长管蚜产生的影响更显著。与单一高温胁迫相比，无论高温、吡虫啉胁迫次序如何，昆虫遭受高温、吡虫啉双重胁迫后均会对生物当代适合度性状产生更严重的负面影响，同时也会产生更加显著的跨代效应，对子代产生显著的影响，既包括负面影响也包括正面刺激作用。

（5）当代高温和吡虫啉胁迫对子代的种群动态产生显著影响。无论是经历单一高温胁迫还是高温、吡虫啉双重胁迫，对麦长管蚜子代的负面影响均集中在发育和存活，如延长子代的发育历期，降低子代的若虫存活率，但这种负面影响一般不会延续到子代的成蚜性状，甚至会对子代的成蚜性状产生正面刺激。尽管母代胁迫经历不会影响子代成蚜性状甚至会产生刺激作用，但是由于胁迫对子代发育的负面影响，对子代总体内禀增长率的贡献更大，因此会导致子代种群增长率整体下降。

（6）早期热背景经历影响后期高温和农药响应表现，甚至产生显著跨代效应。当代若蚜期的高、低强度热背景，在显著提高成蚜期耐热性的同时，也对成蚜的耐药性产生了显著影响。若蚜期低强度热背景经历提高了成蚜耐药性,而高强度热背景经历使成蚜对农药更加敏感。此外,当代若蚜期热历史背景的影响可以延续到子代。有趣的是,对当代成虫繁殖表现产生负面影响的高强度温度背景经历对子代成虫繁殖表现有正面的促进作用,且当代热背景经历也会改变吡虫啉对子代生活史性状的影响,当代单独吡虫啉处理不产生跨代效应,但经历热背景驯化后,吡虫啉的影响就会产生跨代效应,如显著延长子代发育所需时间,从而降低种群增长速率。

在本书的撰写过程中,作者不仅参阅、引用了很多国内外相关文献资料,而且得到了同事亲朋的鼎力相助,在此一并表示衷心的感谢。由于作者水平有限,书中疏漏之处在所难免,恳请同行专家以及广大读者批评指正。

作 者
2021 年 4 月

目 录

第 1 章
绪 论

1.1 高温对昆虫影响研究进展　　4
1.2 农药对昆虫影响研究进展　　7
1.3 高温和农药互作对昆虫影响研究进展　　9
1.4 本书写作背景　　11
1.5 研究目的和意义　　13
1.6 研究创新点　　14

第 2 章
高温类型与吡虫啉互作对麦长管蚜生活史性状的影响

2.1 材料与方法　　18
2.2 结果与分析　　24
2.3 结论与讨论　　58

第 3 章

高温和吡虫啉胁迫次序对麦长管蚜生活史性状的影响

3.1 材料与方法　　　　　　　　　　　68
3.2 结果与分析　　　　　　　　　　　73
3.3 结论与讨论　　　　　　　　　　　94

第 4 章

热背景经历对麦长管蚜高温和农药胁迫响应的影响

4.1 材料与方法　　　　　　　　　　　100
4.2 结果与分析　　　　　　　　　　　104
4.3 结论与讨论　　　　　　　　　　　120

第 5 章

结　论

参考文献

第 1 章

第 1 章

绪 论

一百多年来人类频繁的近代工业活动已造成地球生存环境的显著改变,包括各种污染和全球性环境变迁,如全球气候变暖、臭氧层耗损、化学农药污染等,地球生物面临着比历史上任何时期更强的人为环境选择压力[1,2]。气候变暖被认为是改变生物多样性的最主要人为压力之一[3]。作为气候变暖重要特征的极端高温事件显著增加,不仅影响物种适合度[4]、群落结构[5],甚至限制物种的分布[6]。另一种重要的人为压力就是化学农药的大面积使用[7]。尽管,化学农药在很短时间内可把大面积严重发生的病虫害控制下去,但由于广泛、频繁使用化学农药,导致了病虫害抗药性增加[8],产生了严重的农药残留和食品安全问题[9],此外农药的亚致死剂量也会产生毒物兴奋效应,刺激后代繁殖从而引起病虫害的再猖獗[10]。

昆虫作为地球上种类最多,数量最大的动物群体,几乎遍及整个地球,不但对农业生产以及人类健康产生重要的影响,也在整个生态系统中起着重要的平衡作用。事实上,昆虫的个体发育及生殖等过程都易受到周围环境变化的干扰,如温度、湿度、光照等均会改变昆虫的发育、生殖等基本生活史性状[11]。当前随着气候变暖和化学农药的广泛使用,全球范围内极端温度和接触农药的情况正变得越来越频繁和严酷,因此理解昆虫在气候变暖、化学农药频繁施用背景下的发展趋势,对化学农药风险性评估、生物多样性保护以及农药害虫的综合治理具有重要的意义。

◆ 1.1 高温对昆虫影响研究进展

未来气候变暖,平均气温持续升高的趋势下,高温事件的出现更加频繁、持久[12,13]。全球耦合气候模型表明,欧洲和北美洲地区上空的热浪与一种特定的大气环流模式相吻合,这种改变会导致未来这些地区的热浪天气更加频繁[14]。近50年来中国北方地区,日最高温的增温态势也是十分明显的[15]。与1961～1991年华北地区所统计的逐日气象数据相比,1992～2005年华北平原大部分地区高温天发生频率、强度也显著增加[16]。总之,高温事件频发已经引起了国际广泛关注,成为气候变暖研究中的热点问题之一[17]。

昆虫是典型的变温动物,保持和调节自身温度的能力相对较弱,对外界温度变化的感知极为敏感,受温度变化的影响尤其显著[18-20]。已有的温度对昆虫的研究多涉及的是平均温的变化,事实上,高温事件的日趋增加给昆虫带来的挑战更大,它不仅会改变昆虫原来的生长轨迹(如延长发育历期、减少繁殖等),影响昆虫的神经系统[21]、生殖系统[22]、免疫系统[23]以及生物大分子的合成[24],甚至会造成个体的大量死亡[25]以及种群崩溃[26]。在此背景下,深入研究高温事件对昆虫动态、进化方向的影响就显得非常重要。

1.1.1 高温对昆虫世代内性状的影响

高温直接影响昆虫的存活。根据昆虫对高温事件的耐受性响应,通常可以分为两类:致死高温与亚致死高温。致死高温可在短时间(数小时,甚至数秒)内引起致死效应[25],随着处理温度的升高,产生致死效应所需时间就会缩短。生产上多用作对储物或检疫害虫的物理防治,如大多数储物类害虫在45℃高温下处理12 h,50℃处理5 min以及60℃处理30 s均不能存活[27]。Kalosaka等[28]发现,经45℃热激处理1 h后,地中海果蝇(*Ceratitis capitata*)成虫的死亡率高达100%。当温度达39℃,豚草卷蛾(*Epiblema strenua*)发育期各虫态均不能存活[29]。温室大棚中采取40~48℃高温闷棚2 h,可导致主要害虫烟粉虱(*Bemisia tabaci*)的死亡率在90%以上[30];黄瓜大棚同样采用40℃以上高温处理2 h,对南美斑潜蝇(*Liriomyza huidobrensis*)幼虫、蛹、成虫的防治效果分别可达86.4%、88.2%和92.4%[31]。

而自然环境条件下的高温常为亚致死高温,通常不会引起昆虫大面积急速死亡,但会影响昆虫的耐热性或后期虫态的存活率。Bahrndorff等[32]研究发现,跳虫(*Orchesella cincta*)经亚致死高温35.4℃前处理后,会提高其在37.4℃极端高温的存活率。类似地,苹果蠹蛾(*Cydia pomonella*)成虫经1 h 37℃的热锻炼可以将43℃极端高温存活率从20%显著提升至90%[33]。此外,对黑脉金蝶(*Danaus plexippus*)各阶段发育期进行亚致死高温36℃处理后,不仅会引起当前处理阶段虫态的死亡,也会导致已恢复正常温度的其他发育阶段虫态的死亡[34]。

相较存活,昆虫的生长发育和繁殖对环境高温更为敏感[35-37]。昆虫具有适宜的生存温度范围,在该范围内发育速率会随着温度的升高而加快,但是当温度超过适宜范围就会明显抑制发育。四纹豆象(*Callosobruchus maculatus*)随着温度的升高,整个发育历期先缩短后延长[38]。褐飞虱(*Nilaparvata lugens*)1龄若虫经过亚致死高温41.8℃的热激处理后,显著延长了发育所需的时间,抑制了成虫的繁殖,但却增加了雌虫的寿命[39]。小菜蛾(*Plutella xylostella*)低龄若虫在35℃以上高温下饲养会导致小菜蛾发育畸形出现小型蛾,影响正常交配产卵[40]。再者,对存活影响不大的亚致死高温处理也会显著影响昆虫后续的繁殖性状。例如,Zhang等[41]发现对即时存活率没有显著影响的单一热事件也可导致小菜蛾后续净生殖下降21%。类似的,雄果蝇(*Drosophila buzzatii*)经历的热胁迫温度越高,与其交配的雌蝇的产仔量就会越低[42]。成熟蚜茧蜂(*Aphidius avenae*)经亚致死高温36℃热锻炼1 h后,雌蜂的繁殖量就会下降但寿命却会延长[43]。

此外,高温通过影响昆虫存活、发育、繁殖等基本生命参数,进而也会影响昆虫的种群动态。一方面,适度的温度升高,可加快昆虫的新陈代谢,缩短其世代周期,刺激繁殖,加快种群增长速率。例如,Gomi等[44]通过恒温试验发现,适宜的温度升高会缩短美国白蛾(*Hyphantria cunea*)的发育历期,增加其发生世代数,提升整个种群的增长速率。Maistrello等[45]同样发现,春季温度的升高,茶色缘蝽(*Arocatus melanocephalus*)成虫的生殖力会随之增强,促使其种群密度的升高。另一方面,高温的暴露,也可导致昆虫存活率和繁殖力的显著降低,减缓种群的增长[46,47]。随着高温事件频率和强度的增加,麦蚜的繁殖力显著下降,从而使其种群参数受到明显抑制[5]。

1.1.2 高温对昆虫跨世代性状的影响

高温不仅会影响昆虫自身的生活史性状,也可能对其后代的表型等产生影响[48,49]。近年来,对跨代的表型可塑性的研究,表明子代表型的差异不只取决于基因的遗传,亲代的经历对子代的表型影响也非常重要。这种由于母代生存环境所介导的跨代影响即为母代效应[50,51]。尽管,母代效应的重要性已得到广泛的认可,但是涉及母代效应的研究多关注的是营养、光照、密度、年龄的母代效应[50-55],事实上,温度母代效应所产生的后果也是值得重视的。

亲代高温经历影响后代大小、存活、发育、繁殖等基本的生活史性状,甚至会影响后代的种群动态[56,57]。多数昆虫的雌性成虫在发育或者产卵期经历高温所产的卵体积都较小[58]。例如,双斑湿步甲(*Notiophilus biguttatus*)亲代产卵温度较高所产的卵较小体重较低[59]。类似的,偏瞳敝眼蝶(*Bicyclus anynana*)在较高温度产卵时,卵的体积较小,但却有较高的存活率,从而保证其繁殖的成功率[60]。此外,大量的研究也表明,母代经历的环境变化也会影响后代发育和繁殖等适合度[61,62]。褐飞虱(*Nilaparvata lugens*)无论是母代1龄期还是成虫期受到热激处理,子一代卵的发育历期会显著受到抑制而延长[39]。阿尔蚜茧蜂(*Aphidius ervi*)的亲代在亚致死高温25 ℃和28 ℃分别热激1 h和48 h后,其子代的发育历期、寄生发育完成率均会受到负面影响,但子代的卵孵化率却随着亲代处理温度的增加而增加[63]。黑腹果蝇(*Drosophila melanogaster*)父母饲养的温度越高,他们的子代越具有较高的适合度,这可能与他们的子代发育加快有关[62]。此外,跨代高温暴露的时间越长,对后代的影响越大,通常会导致昆虫种群参数的下降。Guo等[64]研究发现,将烟粉虱(*Bemisia tabaci*)连续4代均暴露于35 ℃高温下,其后代卵的孵化率、存活率会显著降低,从而使得该种群的内禀增长率明显下降。三色书虱(*Liposcelis tricolor*)连续数代暴露于33 ℃亚致死高温下,种群内禀增长率会从第一代的0.042 4降低到第六代的0.023 4[65]。

◆ 1.2 农药对昆虫影响研究进展

目前,对于病虫害的防治最有效、直接的措施仍然是化学农药[66]。农药施用后,在消灭有害生物的同时,也会杀害天敌和其他有益生物,使得物种的多样性显著下降,进而破坏生态平衡[67-69]。有研究表明,在大豆田施用杀虫剂后,中性昆虫、害虫和天敌的多样性、均匀度和丰富度均会产生较大差异,其中中性昆虫受到的影响最大[70]。此外,化学农药的施用除了直接导致靶标生物致死外,随着施药时间的推移,农药本身的理化性质如稳定性和毒力逐渐下降,部分个体不会立即出现中毒或者致死效应[71]。这些幸存的个体与正常个体相比,在行为、生理、生态等方面发生复杂变化,甚至逐渐产生抗药性[72,73]。近年来,有关农药导致的亚致死效应研究越来越受重视[74-76]。

1.2.1 农药对昆虫世代内性状的影响

除了直接的毒杀效应外,农药对害虫世代内的亚致死效应还包括对发育、行为、繁殖等性状的影响。斜纹夜蛾(*Spodoptera litura*)经亚致死剂量的拟除虫菊酯类杀虫剂和虫酰肼处理后,其若虫期以及蛹期均被延长[77]。类似的,小菜蛾(*Plutella xylostella*)幼虫经氯虫苯甲酰胺亚致死剂量处理后,体重明显下降,蛹历期和成虫产卵历期显著延长,但其当代幼虫生长发育历期显著缩短[78]。同时,也有研究表明,昆虫在受到农药刺激后,会出现生态行为的改变。都振宝[79]对荻草谷网蚜取食刺吸式电位仪 EPG 研究发现,在经亚致死剂量的吡虫啉和噻虫嗪处理过的麦苗的取食时间明显低于未经杀虫剂处理过的对照麦苗,说明这两种化学杀虫剂使荻草谷网蚜产生拒食反应。另外,农药对天敌昆虫也可产生显著的亚致死负面效应。Walker 等[80]研究发现,天敌草蜻蛉(*Micromus tasmaniae*)长期取食了经杀虫剂三唑磷处理后的蚜虫,其化蛹率会受到抑制,且幼虫的整个发育历期也会显著延缓。瓢虫 *Coleomegilla maculata* 和 *Hippodamia convergens* 经亚致死浓度的氯虫苯甲酰胺处理后,会导致这两种瓢虫的产卵前期受到显著抑制[81]。捕食性蝽(*Macrolophus pygmaeus*)经噻虫啉亚致死剂量处理后,存活个体休息时间显著增加,捕食行为明显降低[82]。

此外,化学农药对昆虫繁殖性状也有显著影响。一方面,杀虫剂对昆虫繁殖具有抑制作用。Bao 等[83]发现,与未经农药处理的对照相比,经亚致死剂量吡虫啉和呋虫胺处理不同翅型的褐飞虱,繁殖力均显著被抑制,长翅型褐飞虱分别下降了 68.8% 和 52.4%,短翅型下降了 57.9% 和 43.1%。化学农药对天敌的生殖力也有显著影响。多异瓢虫(*Hippodamia variegate*)接触亚致死浓度的阿维菌素、高效氯氟氰菊酯以及吡虫啉后,每日的平均产卵量显著下降,总的产卵量也下降明显[84]。类似的,将小花蝽(*Orius laevigatus*)成虫饲养在经多杀菌素和阿维菌素亚致死浓度浸渍后的马铃薯叶片上 1 h,不仅当代雌虫的繁殖力显著下降,而且后代的生殖能力也会明显受到抑制[85]。另外,昆虫的某些性状对化学农药也可能可能会产生毒物兴奋作用。Cho 等[86]等发现,经亚致死剂量的氟啶虫酰胺或噻虫嗪处理的桃蚜(*Myzus persicae*),寿命显著延长,增加能力显著增加。Fariba 等[87]同样发现,蚜小蜂(*Encarsia inaron*)幼虫经亚致死剂量的吡虫啉处理后,产卵量等繁殖性状、内禀增长率等种群参数显著高于对照。

1.2.2 农药对昆虫跨世代性状的影响

同样的,化学农药不仅会对昆虫当代的生命表参数产生影响,也可能通过母代效应影响其后代的表型。惠婧婧等[88]通过研究亚致死剂量吡虫啉对豌豆蚜(*Acyrthosiphon pisum*)的影响发现,随着接触吡虫啉剂量的增加,在当代寿命会明显缩短,繁殖力降低的同时,子一代的发育历期也显著延长,产蚜量显著减少,种群参数显著降低。2龄若虫小菜蛾(*Plutella xylostella*)经高效氯氟氰菊酯和茚虫威亚致死剂量处理后,不仅当代小菜蛾的生长发育、繁殖显著受到抑制,这种负面效应会持续到下一代,抑制后代的生长发育以及种群增长[89]。Yin等[90]也发现,用亚致死剂量的多杀菌素处理小菜蛾3龄幼虫会显著影响当代和下一代的生长、发育以及繁殖,从而对该种昆虫的种群动态产生负面影响。但也有文献表明,烯啶虫胺亚致死剂量处理4龄棉蚜(*Aphis gossypii*)导致当代成蚜寿命和繁殖显著下降,但却对F1代的寿命和繁殖产生毒物兴奋效应存在刺激生殖的现象,进而加快了种群的生长速度[91]。

农药的亚致死效应还会影响后代的性别比例、翅型分化,如红切叶蜂(*Osmia bicornis*)接触亚致死剂量的新烟碱类农药,显著降低了当代的繁殖力,但却导致其子代雄虫性别比例显著增加[92]。Wang等[93]通过研究6种杀虫剂的亚致死剂量对桃蚜(*Myzus persicae*)翅型分化的影响发现,在试验室中,桃蚜接触亚致死剂量的吡虫啉和戊氰菊酯后,其后代中有翅蚜的比例明显高于对照;而在温温室条件下,除了经吡虫啉和戊氰菊酯处理的桃蚜后代的有翅比例明显增加外,阿维菌素也显著增加了后代中有翅蚜的比例。类似的,Conway等[94]也曾报道经吡虫啉处理后的棉蚜,会诱导其后代产生更多的有翅蚜,从而刺激棉蚜的扩散。

◆ 1.3 高温和农药互作对昆虫影响研究进展

在自然界中,环境高温和化学农药对昆虫的影响不是独立存在的,两者之间有着复杂的交互影响。玉米螟(*Ostrinia nubilalis*)在较高温度下对某些农药的敏感性降低,温度升高预期会增加这些农药使用量[95]。较高温度下,昆虫本身较快的热适应能力、较高的农药降解速率[96]以及发育速率加快[35]造成的较短农药接触时间均可能降低农药的负面影响。高温也可以加速昆虫代谢,增加农药摄入量[97],提高农药的毒性[98],分耗能量用以抵抗高温从而降低对化学农药的能量分配,增加昆虫对农药的敏感性[99],加重对昆虫的负影响。反过来,农药也会增加物种对气候变暖的敏感性,削弱物种响应气候变暖及其相关高温事件的能力[100]。同时,长期、低水平农药暴露也可能导致物种的进化权衡,使其适应气候胁迫的能力下降[101]。两者之间的这种错综关系,预示着高温和农药交互胁迫对昆虫影响的复杂性和挑战性。

温度对化学农药自身理化性质及生物的新陈代谢等的影响,成为影响化学农药活性的重要因素之一[102]。具体而言,温度不仅会影响昆虫的行为活动[103]、解毒代谢[104,105],也会影响农药的物理化学性质如挥发性、稳定性和降解的速率等[96,106],从而改变靶标害虫对于化学农药的敏感性。温度对化学农药的即时毒力效应一般用温度系数来表示,通过测定系列恒温梯度下化学农药对昆虫的毒力变化来获得[107]。已有研究证实:不同作用机制化学农药的温度系数表现并不相同,同类化学农药对不同种昆虫,甚至同类化学农药不同药剂品种对同种昆虫的温度效应也可能存在较大差异[108,109]。一般认为,有机磷类化学农药和氨基甲酸酯类化学农药的毒性与温度呈正相关,即正温度效应[110,111],拟除虫菊酯类化学农药多呈负温度系数[112,113],新型化学农药中的新烟碱类、吡唑类等化学农药常表现为正温度系数[114,115]。

大量研究关注于对高温和化学农药互作对昆虫耐热、耐药性的影响。通常情况下,热处理与化学农药可诱导多种昆虫产生交互耐受性,对机体产生保护作用。例如,亚致死农药前处理增加褐飞虱耐高温能力,提升该物种在高温暴露下的存活率[116,117],热激前处理增加烟粉虱[118]和斯氏按蚊(*Subgenus stephensi*)[119]的耐药性,而药后高温处理也可增加玉米螟抗药性[95],两种胁迫条件均可诱导斯氏按蚊、埃及伊蚊(*Aedes aegypti*)[120]和小菜蛾(*Plutella xylostella*)[121]产生交互耐受性。热激蛋白应激反应[116,122,123]和解毒酶活性变化[109,118]可能介入了热激以及化学农药诱导抗性的生理机制。与高温或化学农药单作相比,两种胁迫联用诱导绿盲蝽(*Apolygus lucorum*)产生了更高的 Hsp90 表达量[124]。但热激并没能诱导马铃薯幼虫(*Leptinotarsa decemlineata*)对吡虫啉产生预期的 Hsp70 应激响应[125]。

少量试验模拟开展了生物响应高温与化学农药互作的生态性状研究,但是主要围绕水生动物展开,重点关注水体环境生态安全评价及生物多样性保护[97,101]。涉及昆虫的研究主要集中于早期具有水生(卵、稚虫期)生活史的蜻蜓目豆娘和双翅目蚊类等。一方面,高温和农药双重胁迫可产生直接的即时互作生态效应。二者的联合作用可降低库蚊(*Culex restuans*)和白纹伊蚊(*Aedes albopictus*)的存活[126];吡虫啉、阿维菌素、抗蚜威以及印楝素的亚致死剂量处理后会加速桃蚜(*Myzus persicae*)在高温下的死亡[127];温度升高可以降低农药对豆娘(*Ischnura elegans*)致死率[96,128]。另一方面,高温和农药还可产生延迟互作生态效应。例如若虫期三唑磷暴露可提高褐飞虱(*Nilaparvata. lugens*)高温成虫的繁殖能力[129,130],豆娘卵期高温可增加稚虫期农药暴露的致死率[131],反过来稚虫期农药暴露也可降低热激成虫的热激蛋白响应,加速免疫物质的下降[132]。

一些研究也涉及了跨代影响。例如,西花蓟马成虫期热激与农药双重胁迫可产生显著的当代即时和跨代延迟负面效应,当代成虫死亡率和氧化损伤增加[133],F1 代若虫和蛹发育延长、成虫寿命和繁殖力降低[134]。表明高温和农药的互作后果相当复杂,需要更多的研究证据加以揭示。

◆ 1.4 本书写作背景

如前所述,尽管热药互作对昆虫的影响已经引起了广泛关注,但是大多数研究通常忽视了以下几个方面。

(1)不同类型亚致死高温和农药互作产生的影响。一方面,大多数研究温度为梯度恒温,不能反映自然变温热事件的生理生态效应[135-137];另一方面,已有热事件多涉及同一高温强度不同作用时间[41,138]或同一作用时间不同高温强度[139,140]对生物的影响,但这类处理方式对生物所施加的胁迫呈递增趋势,胁迫程度越高,对生物产生的负面影响越大,产生的后果可以预测。此外,自然田间很少出现高温致死的现象。这是由于昆虫对于高温事件并不仅仅是被动接受,也会表现为主动地避热行为[141,142],且昆虫响应不同强度高温的行为热调节时间也存在显著差异[103,143]。不同亚致死高温强度与行为反应时间组合是否会对昆虫生活史性状产生显著影响,特别是当这类高温与化学农药相结合时会产生什么样的互作效应,是比高温致死效应更值得我们关注的问题。

(2)高温与化学农药出现次序的重要性。由于高温[118]和农药[116]均能诱导昆虫产生生理应激响应,从而对另一种胁迫产生交互耐受(敏感性)。考虑到两种胁迫性质上的极大不同,昆虫的应激响应机制也可能存在显著差异。在此背景下,高温与农药两种胁迫出现次序就显得尤为重要,但这个关键因素却几乎没有受到任何关注[97]。高温与农药的出现次序通常有三种情况:先药后热、热药同时和先热后药。但生产中为减少农药中毒事故的发生,常常避开中午的高温而在早晨或黄昏用药,因此开展先药后热、先热后药两个方向上互作效应的研究具有非常重要的现实意义。

(3)高温与化学农药双重胁迫可能的跨代效应。已有的高温和农药结合研究,通常关注的是二者对即时存活以及世代内延迟效应的影响,鲜有研究关注跨代的影响。事实上,生存环境可能通过母亲的资源分配从而对子代的表现产生影响,产生典型的母代效应[48,144]。

(4)热背景经历对于昆虫响应农药胁迫的影响。首先,自然热事件通常以不同频率间歇高温的形式出现,且高温天气有时甚至会持续数天。因此,一个生物的表现反映的是其一生所经历的栖息地环境变化总和的影响。某个基因型表型性状值的变化不但取决于当前环境,还取决于过去环境的时间滞后效应[145]。在气候变暖、极端热事件增加的大背景下,生物的热背景经历对其应对其他环境胁迫的影响应受到更多关注。

本书在充分考虑热事件特征、胁迫次序以及热背景经历的基础上,以中国北方麦区小麦蚜虫主要优势种麦长管蚜(*Sitobion avenae*)为研究对象,以防治刺吸式口器害虫的常用新烟碱类化学农药吡虫啉作为试验药剂,通过两种环境胁迫互作对麦蚜核心生命参数和种群适合度的影响研究,阐明麦蚜响应气候变暖和化学农药交互胁迫的生态表现。在此基础上,通过对麦蚜跨代效应的研究,明确后代对母代环境效应的继承以及响应能力,同时揭示热历史背景对生物响应农药胁迫的影响,合理评估麦蚜响应气候变暖热事件和化学农药交互胁迫的生活史表现,预测麦长管蚜的种群命运。

1.5 研究目的和意义

生物如何应对两大人为压力——气候变暖和化学农药的挑战已成为国际热点问题。全球气候变暖背景下,作为化学农药靶标的农业害虫更易遭遇高温热事件和化学农药的双重胁迫。尽管热激或恒温条件下昆虫对两种胁迫的交互耐受性研究取得一定进展,但生态背景下两种胁迫的综合效应尚不清晰。因此,开展深入的、基于两种环境变化背景下的昆虫生态学研究,理解二者交互胁迫对地球生物,尤其是对农业重要害虫生态性状的潜在影响,对气候变化情景下的生物多样性保护和害虫综合治理有着重要的理论价值。

麦长管蚜是全球重要的麦类害虫之一,同样也是我国麦区头号害虫,百株蚜量多达450头,严重受害田块甚至在千头以上,给麦类作物造成了严重的产量、质量损失[146]。麦长管蚜体形小,身体的比表面积大,与外界环境对地热传导和热交流较快,对外界温度变化非常敏感[147,148]。蚜虫特有的套叠世代[149],意味着该物种体内胚胎还包含着正在发育的胚胎,会产生较强的世代内延迟效应和世代间母代效应。与此同时,麦长管蚜生活周期短、繁殖能力强,因此其爆发频率较高[150,151]。目前防治麦蚜的主要手段仍然是化学农药,而吡虫啉作为近几年发展较快的新烟碱类杀虫剂,广泛用于防治刺吸式口器的害虫[152,153]。因此,开展麦蚜响应高温和化学农药交互胁迫的生态表现,对气候变暖、化学农药广泛应用背景下的麦蚜抗性治理及综合防控也具有重要的现实生产意义。

◆ 1.6 研究创新点

(1)将符合自然变化规律的不同类型的亚致死高温事件而非恒温或恒定变温纳入研究体系,在研究方法上具有特色。如前所述,以往研究涉及恒温或恒定变温温度设计与自然条件下的高温事件特征存在很大出入,由此推断高温事件和化学农药互作导致的自然生态后果时可能会出现很大偏差。本书在充分考虑昆虫对高温响应的短时自主行为热调节的基础上,设计了 3 种高温强度组合行为反应时间的亚致死热事件(34 ℃/180 min、36 ℃/30 min 或 38 ℃/10 min)。设计了符合自然变化特点的不同高温时间类型和高温事件频率且同时一定程度上达到了以有限试验研究代表繁杂多变自然气候特征的目的。在气候变化生态学研究方法上具有特色。

(2)阐明生态相关热事件与化学农药相互胁迫下的昆虫生态响应,在研究思路上有创新之处。以往相关研究忽视两胁迫次序、经历胁迫后的跨代效应以及生物的热背景经历等关键环境条件对昆虫表型可塑性的重要影响。本书通过贴近自然的模拟试验设计,将以上生态相关关键环境因子有机地融入"热事件—化学农药—麦长管蚜"研究体系,通过对重要表型性状:当代和跨代生活史性状及种群适合度(直观生态表型)的影响研究,揭示气候变暖热事件频发、化学农药广泛使用背景下昆虫的生态表现响应机制。在气候变化生物学和昆虫毒理学研究思路上具有创新之处。

第 2 章

第 2 章

高温类型与吡虫啉互作对麦长管蚜生活史性状的影响

近年来,随着全球气候变暖,高温事件出现得越发频繁[14,154],热浪天气持续的时间也逐渐增加[155]。春夏季农作物及农业害虫的主要生长季,高温天气发生尤其频繁[156],有时甚至会持续数天[157]。但昆虫对于高温事件并不仅仅是被动接受,也会表现为主动热调节行为[141,158]。例如当昆虫感受到极端高温时,在短时间内就会做出避热反应,逃至阴凉的地方,以躲避高温的伤害[141,142,159],因此田间很少出现高温致死的现象。且昆虫响应不同高温的行为热调节反应时间也存在显著差异[103,143]。这类持续一到数天且未导致明显死亡的高温,特别是不同强度高温与行为热调节反应时间组合是否会对昆虫世代内及跨代生活史性状产生显著影响,尤其是当这种高温与化学农药相结合时会产生什么样的互作效应,是比高温致死效应更值得我们关注的问题。

麦长管蚜是我国小麦种植区的重要害虫,目前我国对于小麦害虫的防治仍然以化学农药为主,而过去常常使用的有机磷类、拟除虫菊酯类等常见农药,由于大量、连续使用,已引起麦蚜产生不同程度的抗药性,防效减弱[160,161]。但近年来,吡虫啉以其良好的防治效果已成了防治各种刺吸式害虫的首选农药品种。考虑到麦长管蚜的防治是在自然环境温度不断变化的条件下进行的,所以麦长管蚜在发生期间可能会同时受到不同类型的高温和农药交互所带来影响,据此,我们提出如下问题:

（1）持续一至数天的高温强度、行为反应时间组合与农药互作是否会对麦长管蚜的即时存活及后续延迟生活史性状产生显著影响？如果有互作效应,不同的生活史性状是否会产生不同的响应？

（2）持续一至数天的高温强度、行为反应时间组合与农药互作是否会产生显著跨代延迟生态效应？如果有,这种影响会持续多久,主要影响后代的哪些性状？这些影响是否会因热事件类型的不同而产生差异？

为了回答上述问题,我们考察了1 d、3 d或5 d的三种高温强度、行为反应时间组合(34 ℃/180 min、36 ℃/30 min或38 ℃/10 min)与亚致死剂量的吡虫啉(LC_{20}=20 μg/mL)互作对麦长管蚜成蚜当代即时存活率的联合作用,以及对寿命、繁殖、种群参数和子一代生长发育、繁殖、种群参数等生活史性状的延迟影响。

◆ 2.1 材料与方法

2.1.1 供试虫源

试验所用麦长管蚜于2016年5月采自临汾(35°55'N,111°16'E),在养虫室内连续用5～20 cm高的小麦幼苗在养虫笼(60 cm×60 cm×60 cm)内饲养备用。每周更换1次种植在营养钵(12 cm×10 cm)内的新鲜小麦幼苗。饲养条件:养虫室温度为(22±0.5)℃,湿度为50%～70%,光周期(L:D=16:8)。之后各章试验用虫来源相同。

2.1.2 试验用虫

挑取饲养条件下的新生若蚜约 4 000 头,用毛笔轻轻接入新鲜小麦幼苗上集中饲养,大约 200 头/钵、3 钵(养虫笼),置于养虫室内集中饲养至 9 日龄用于试验。试验开始前,将 9 日龄麦长管蚜于饲养管中独立分装。饲养管由中空的玻璃管(直径 15 mm,高 70 mm)和上下端塞有透气海绵塞(直径 17 mm,高 20 mm)组成。使用时,将底部海绵浸水,中间插入麦叶(高度 50 mm),塞入饲养管底部,将单头蚜虫用毛笔接到管内麦叶上,之后用干燥海绵塞将饲养管上部塞住。

2.1.3 供试药剂

供试药剂为 95% 吡虫啉原药(中农联合生物科技有限公司提供)。

2.1.4 试验因子设定

2.1.4.1 高温事件

根据预试验,分别记录置于 34℃、36℃、38℃ 恒温下 9 日龄麦长管蚜由不动或者缓慢移动突然开始剧烈活动的平均时间,该时间主要考虑蚜虫感到环境不适时会产生自主的避热行为,最终确定高温强度、行为反应时间组合设置:34℃/180 min、36℃/30 min、38℃/10 min 三种类型。设置热处理持续天数为 1 d、3 d、5 d,模拟自然界高温天 1 d 到数天的持续时间[41]。高温事件设计为 9 种:3 种亚致死高温类型×3 种持续天数。

其中,亚致死高温处理模式以 24 h 为周期,34℃、36℃、38℃ 热处理分别在 3 台恒温 34℃、36℃、38℃ 人工气候箱(宁波江南仪器厂,RXZ-280B)中进行。每天分别经历 34℃/180 min、36℃/30 min、38℃/10 min 高温处理,同时为避免夜间低温变化的潜在效应[162],一天中剩余时间设置为 22℃标准恒温,由一台人工气候箱控制。所有人工气候箱的箱内湿度保持 40%~60%,光周期(L:D=16:8)。

2.1.4.2 毒苗制备

为了确定施药浓度,采用浸叶法[72]对 9 日龄蚜虫进行毒力测定,方法略有改动。将吡虫啉原药用比例为 1∶9 的吐温 80(Tween 80)和丙酮分析纯作为溶剂制备成 10 mg/mL 的原液,再用蒸馏水稀释成 5 个系列质量浓度,以蒸馏水作为空白对照,之后将麦苗浸到这一系列质量浓度中 5～7 s,在空气中晾干,放入一个由滤纸保湿的培养皿中(直径 90 mm)10 片毒麦苗/皿,接入 30 头 9 日龄麦长管蚜,24 h 后记录存活蚜虫的个数。每处理重复 3 次。采用机值分析法进行数据分析(SPSS 19.0)。获得毒力回归方程:$y=-1.917x+0.823$,选取亚致死剂量 $LC_{20}=20$ μg/mL 为试验农药处理质量浓度,试验时,将长度约为 50 mm 的麦苗在该浓度浸没 5～7 s 后取出,晾干,作为毒苗。

2.1.5 试验设计

2.1.5.1 高温和吡虫啉对麦长管蚜即时存活率联合作用测定

开展亚致死高温类型与农药吡虫啉互作对麦长管蚜存活的联合作用测定,其中高温处理周期为 1 d、3 d 和 5 d,农药吡虫啉处理时间为 1 d(图 2-1)。编号 H_1 代表高温单作(H 为 heat,1 为高温处理持续 1 d),H+P_1 代表高温农药同时处理(H 为 heat,P 为 pesticide,1 为高温处理持续天数 1 d)。整个试验由 9 个高温单作(3 种高温类型 ×3 种持续天数)+9 个高温和农药互作+1 个 22℃农药单作(P)+1 个 22℃空白对照(CK)=20 个处理。

高温+农药互作处理(H+P):9 日龄麦长管蚜,每 10 头置 1 支 1.5 mL 离心管中,管中放有毒苗,离心管插入带孔泡沫板固定后,置于培养箱中先进行 9 种高温事件处理。

高温处理结束后,转移到恒温 22℃下继续进行农药处理,1 d 后,调查高温事件+农药死亡数,3 个 1 d 高温处理的试验终止;需要进行 3 d 和 5 d 高温的其余 6 个处理,转移到新鲜麦苗的离心管中,置于 22℃下,按试验温度设计继续完成相应的高温处理,并同样与 1 d 后进行死亡率调查。试验期间麦苗每 3 天更换一次。

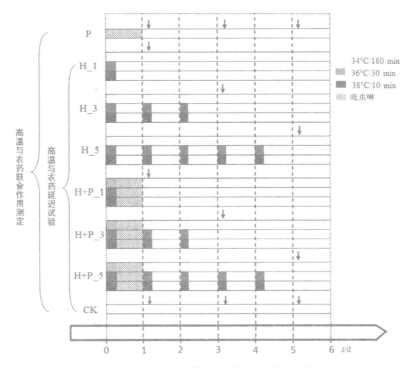

图 2-1 高温与农药吡虫啉互作试验设计图

注:"↓"代表每处理的死亡率调查点以及子代采样点。空白对照子代采样点为第一个"↓"代表的时刻。第 0 天被定义为试验开始的时间。

Fig.2-1 Experimental design of heat events interaction with imidacloprid.

Note: ↓ indicates mortality assessment and offsprings sampling points for each treatment. The first ↓ indicate offspring sampling point for CK. "0" indicated when the experiment started.

高温单作(H):9 日龄麦长管蚜,每 10 头置于 1 支 1.5 mL 离心管中,管中放有新鲜麦苗,离心管插入带孔泡沫板固定后,置于培养箱中进行 9 种高温事件处理,处理结束后 1 d 调查死亡数。试验期间麦苗每 3 天更换一次。

农药单作(P):9 日龄麦长管蚜,每 10 头置于 1 支 1.5 mL 离心管中,管中放有毒苗,离心管插入带孔泡沫板固定后,置于恒温 22℃下,1 d 后调查死亡数,作为 H+P_1 处理单独药后的死亡率。然后接无毒苗,置于 22℃下,继续调查第 4 天和 6 天的死亡率,分别作为 H+P_3、H+P_5 单独药后死亡率。试验期间麦苗每

3天更换一次。

空白对照(CK)：9日龄麦长管蚜,每10头置于1支1.5 mL离心管中,管中放有新鲜麦苗,置于恒温22℃下,分别于第2天、第4天、第6天调查死亡数,作为不同处理分别持续1 d、3 d、5 d后的空白对照。试验期间麦苗每3天更换一次。

以上所有处理均设置4次重复,每重复10头蚜虫。

2.1.5.2 高温与吡虫啉互作对麦长管蚜生态性状的影响

为了研究亚致死高温类型与农药吡虫啉互作对麦长管蚜当代以及跨代生活史性状的延迟效应,设计了亚致死高温类型、持续天数与化学农药互作的全因子试验。其包括3个高温类型(34℃/180 min、36℃/30 min、38℃/10 min),3个高温持续天数(1 d、3 d、5 d),2个农药处理(有或无吡虫啉处理),同时增加了1个空白对照用于对比高温和农药对生活史性状的实际影响,共计19个处理(图2-1)。其中40头蚜虫/处理用于母代生活史性状调查；另80头蚜虫/处理仅用于子一代采集。

高温单作(H)：将9日龄麦长管蚜单头接入插有新鲜无药麦苗的饲养管后,置于培养箱中9种高温事件处理,处理结束后转移到恒温22℃下继续饲养,试验期间麦苗每3天更换一次。

热药互作(H+P)：将9日龄麦长管蚜单头接入插有浸毒麦苗的饲养管中,置于培养箱中先进行9种高温事件处理,高温处理结束后,转移到恒温22℃下继续进行农药处理1 d后,3个1 d高温处理结束,之后转移到恒温22℃下继续饲养,直至供试蚜虫全部死亡。需要进行3 d和5 d高温的其余6个处理,转移到新鲜麦苗的离心管中,置于22℃下,按试验温度设计继续完成相应的高温处理,之后仍转移到恒温22℃下继续饲养,直至供试蚜虫全部死亡。试验期间麦苗每3天更换一次。

空白对照(CK)：将9日龄麦长管蚜单头接入插有新鲜无药麦苗的饲养管后,置于恒温22℃下一直饲养,直至供试蚜虫全部死亡。

母代性状调查：试验期间,每天都调查记录1次存活状态、产仔情况,并将计数后若蚜去除,直至供试蚜虫全部死亡,母代性状调查结束。存活寿命是从第9日开始算起,直到死亡的时间;繁殖同样从第9日开始记录,整个存活时期每头蚜虫所产后代的总数。

跨代性状调查:所有处理结束后第二天调查时,随机采集30头子代/处理(共18个处理)新生若蚜置于常温22℃养虫室内,于每日上午8点记录死亡个体、蜕皮及产仔情况,并将蜕皮、死亡蚜虫及新生若蚜去除,直至供试蚜虫全部死亡。去除逃逸蚜虫,所有处理蚜虫均测定了以下指标:若虫死亡率、发育历期、成蚜产仔前期、成蚜寿命以及繁殖。若虫死亡率是指活到成蚜的若虫占全部测试若虫的比例。发育历期是指新生若蚜发育至成蚜所需的时间。成蚜产仔前期是指新蜕变的成蚜至成蚜产仔所需时间。成蚜寿命是指从变成成蚜到蚜虫死亡的时间。成蚜繁殖是指每头成蚜所产后代的总数。

2.1.6 统计分析

协同毒力指数 = (校正死亡率 − 理论死亡率)/理论死亡率 × 100;其中,协同毒力指数 ≥ 20 为协同作用,协同毒力指数 ≤ −20 为拮抗作用,处于二者之间为相加作用。

其中:

理论死亡率 = 单独热校正后的死亡率 + 单独药校正后的死亡率 − 单独热校正后的死亡率 × 单独药校正后的死亡率

校正死亡率 = (未处理对照组生存率 − 处理组生存率)/未处理对照组生存率 × 100

母代繁殖、寿命、繁殖率以及子代发育历期和产仔前期性状,在分析之前,需经平方根转换后提高其正态性。转换后,数据统计分析均使用转换后的,但作图为了直观仍用原始数据。母代繁殖、寿命、繁殖率和子代发育历期、产仔前期、成蚜寿命、成蚜繁殖性状均采用一般线性模型(GLM)进行方差分析,利用 Duncan-t 检验法进行高温和持续天数处理间多重比较,利用独立样本 T 检验来比较农药处理间的差异显著性(SPSS 19.0);子代若蚜存活率采用列联表法进行存活率显著性分析,非独立 2×2 以及 2×3 表法实现处理间的多重比较。

为分析母代种群参数,将前期22℃下的9日龄前的存活率统一为100%。而子代种群参数根据实际记录计算。

净增长率 $R_0=\sum l_x \times m_x$,

平均世代时间 $G=\sum l_x \times m_x \times X/\sum l_x \times m_x$,

内禀增长率 $r_m=\ln(R_0)/G$,

其中,"x"为蚜虫的年龄;"l_x"为蚜虫在"x"岁时的成活率;"m_x"代表蚜虫在"x"岁时所产的后代数量。在R语言中采用bootstrap程序计算每一种群参数的均值和相应95%置信区间,并调用R语言中agricolae包中的"sample"函数和"kruskal"函数,采用Kruskal-Wallis法来比较各处理相应种群参数的差异显著性。

◆ 2.2 结果与分析

2.2.1 高温和吡虫啉对即时死亡率联合作用评价

由表2-1可知,不同热事件类型和吡虫啉互作对麦长管蚜所产生的联合毒力结果不同。在高温34℃/180 min下,无论持续1 d、3 d还是5 d与吡虫啉联合毒力表现的均是相加作用;在高温36℃/30 min下,持续3 d与吡虫啉联合毒力表现的是协同作用,但持续1 d、5 d与吡虫啉联合毒力却表现为相加作用;在高温38℃/10 min下,持续1 d、3 d与吡虫啉联合毒力表现的是拮抗作用,但持续5 d与吡虫啉联合毒力却表现为相加作用。

表 2-1 高温和吡虫啉对麦长管蚜即时死亡率联合作用评价

Table 2-1 Assessment of the effects of heat events, imidacloprid and their interaction on immediate mortality of *Sitobion avenae*

处理 Treatment	温度/℃ Temperature	热持续时间/d Duration time /d	第一天吡虫啉处理 Imidacloprid (mg·L^{-1})	调查时间 Time	死亡率 Mortality 实际值 Real	校正值 Corrected	理论值 Theoretical	协同毒力指数 Cooperative virulence index (c.f.)	结果评定 Results assessment
CK	22	0	0	2nd	0.03	0.03			
				4th	0.13	0.13			
				6th	0.30	0.30			
P	22	0	20	2nd	0.08	0.05			
				4th	0.15	0.03			
				6th	0.33	0.04			
H	34	1	0	2nd	0.33	0.31			
	34	3	0	4th	0.68	0.63			
	34	5	0	6th	0.88	0.82			
	36	1	0	2nd	0.15	0.13			
	36	3	0	4th	0.70	0.66			
	36	5	0	6th	0.85	0.79			
	38	1	0	2nd	0.18	0.15			
	38	3	0	4th	0.78	0.74			
	38	5	0	6th	0.88	0.82			

续表

处理 Treatment	温度/℃ Temperature	热持续时间/d Duration time /d	第一天吡虫啉处理 Imidacoprid (mg·L⁻¹)	调查时间 Time	死亡率 Mortality 实际值 Real	死亡率 Mortality 校正值 Corrected	死亡率 Mortality 理论值 Theoretical	协同毒力指数 Cooperative virulence index (c.f.)	结果评定 Results assessment
H+P	34	1	20	2nd	0.30	0.28	0.34	-17.82	相加作用
	34	3	20	4th	0.75	0.71	0.64	11.75	相加作用
	34	5	20	6th	0.90	0.86	0.83	3.54	相加作用
	36	1	20	2nd	0.23	0.21	0.17	18.63	相加作用
	36	3	20	4th	0.85	0.83	0.67	24.24	协同作用
	36	5	20	6th	0.95	0.93	0.79	17.04	相加作用
	38	1	20	2nd	0.08	0.05	0.20	-74.00	拮抗作用
	38	3	20	4th	0.50	0.43	0.75	-42.87	拮抗作用
	38	5	20	6th	0.88	0.82	0.83	-0.77	相加作用

2.2.2 高温与吡虫啉互作对当代生活史性状的影响

2.2.2.1 寿命

高温类型、持续天数和吡虫啉对麦长管蚜寿命的三因子方差分析结果表明（表 2-2）：高温类型、持续天数以及吡虫啉对麦长管蚜当代寿命影响差异显著。高温类型与持续天数以及高温类型与吡虫啉对麦长管蚜当代寿命存在显著的交互作用，但持续天数与吡虫啉以及高温类型、持续天数和吡虫啉对该性状却无交互影响。

表 2-2　高温与吡虫啉互作对麦长管蚜当代寿命的方差分析

Table 2-2 Results of variance analysis for effects of heat events, imidacloprid and their interaction on longevity of *S.avenae*

来源 Source	DF	MS	F	P
高温类型 Heat treatment	2	4.64	3.51	0.030
持续天数 Duration	2	18.25	13.81	<0.001
吡虫啉 Imidacoprid	1	21.02	15.89	<0.001
高温 × 持续天数 Heat treatment × Duration	4	4.40	3.32	0.010
高温 × 吡虫啉 Heat treatment × Imidacoprid	2	9.81	7.42	0.001
持续天数 × 吡虫啉 Duration × Imidacoprid	2	1.55	1.17	0.310
高温 × 持续天数 × 吡虫啉 Heat treatment × Duration × Imidacoprid	4	2.29	1.73	0.141
误差 Error	702	1.32		

与空白对照寿命（9.63±0.98）d 相比，所有处理均显著缩短了当代麦长管蚜的寿命，最少缩短了（2.58±1.42）d（图 2-2）。

无吡虫啉处理的情况下，三种高温类型（34℃/180 min，36℃/30 min，38℃/10 min）仅在持续 1 d 后对麦长管蚜当代寿命影响有显著差异（$F_{2,117}=3.519, P=0.033$），且 34℃/180 min 产生的负面影响最大，麦长管蚜的存活时间仅为（3.48±0.42）d；但在持续 3 d（$F_{2,117}=1.167, P=0.315$）、5 d（$F_{2,117}=1.533, P=0.220$）差异消失 [图 2-2(A)]。

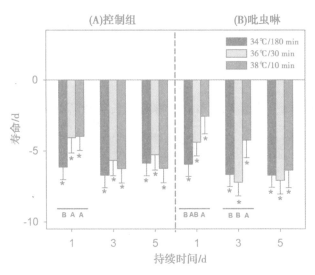

图 2-2 两种农药处理下,3 种高温类型对麦长管蚜当代寿命的影响(平均值 ± 标准误)。每个柱子表示不同处理组减去空白对照组对应的平均值与标准误。"*"表示处理组和空白对照组间差异达显著水平($P=0.05$)。大写字母代表不同色柱高温类型间的差异显著性($P=0.05$)。

Fig. 2-2 Effect of three heat wave types on maternal longevity (mean ± SE) of S.avenae under pesticide treatment (absent versus present). Each bar represents the mean value and standard error corresponding to the different treatment minus the CK treatment. The "*" represent significant difference ($P=0.05$) between treatment and CK group. Different letters indicate significant level ($P=0.05$) of values between different heat wave types.

在吡虫啉存在的条件下,三种高温类型在持续 1 d ($F_{2,117}$ = 4.108, P = 0.019)、3 d ($F_{2,117}$ = 7.529, P = 0.001)后对麦长管蚜当代寿命影响有显著差异,且在持续 1 d 后,34℃/180 min 产生的负面影响最大,麦长管蚜的存活时间仅为(3.68 ± 0.55)d;在持续 3 d 后,(36℃/30 min)产生的负面影响最大,麦长管蚜的存活时间仅为(2.40 ± 0.22)d;在持续 5 d($F_{2,117}$ = 1.646, P = 0.197)后,三种高温类型对寿命无显著影响 [图 2-2(B)]。

在高温类型 34℃/180 min 下,单独高温持续天数 [图 2-3(A) 灰柱,$F_{2,117}$=1.022,P=0.363] 以及高温天数和吡虫啉同时作用 [图 2-3(A) 黑柱,$F_{2,117}$=0.752,P=0.473] 均没有对麦长管蚜当代

寿命产生显著影响；且在该类型下，吡虫啉的有无在高温持续 1 d（$t=-0.179, df=78, P=0.859$）、3 d（$t=-0.096, df=78, P=0.924$）、5 d（$t=1.210, df=78, P=0.230$）后对当代寿命均无显著影响[图 2-3(A)]。

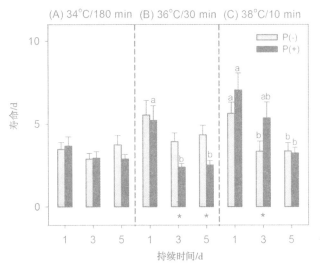

图 2-3 3 种高温类型处理下，高温持续时间和吡虫啉对麦长管蚜当代寿命的影响（平均值 ± 标准误）。"*"表示不同色柱吡虫啉处理间差异达显著水平（$P=0.05$）。小写字母代表同一色柱高温持续天数间的差异显著性（$P=0.05$）。

Fig. 2-3 Maternal longevity（mean ± SE）of S.avenae at different heat wave duration interaction with imidacloprid. "*" indicate significance level（$P=0.05$）of values between pesticide treatment. Different letters indicate significant level（$P=0.05$）of values between different heat wave duration.

高温类型 36℃ /30 min 下，单独高温持续天数[图 2-3(B) 灰柱，$F_{2,117}=1.489, P=0.230$]对麦长管蚜当代寿命无显著影响，但高温天数和吡虫啉同时作用[图 2-3(B) 黑柱，$F_{2,117}=8.551, P<0.001$]却对麦长管蚜当代寿命产生显著影响，高温持续 1 d 和吡虫啉同时作用后，麦长管蚜的寿命为（5.23 ± 0.88）d，显著高于高温持续 3 d、5 d 和吡虫啉同时作用后麦长管蚜（2.40 ± 0.22）d 和（2.52 ± 0.26）d；且在该类型下，无论高温处理持续几天，吡虫

啉的存在对麦长管蚜的寿命均产生了负面影响,且在持续3 d ($t=2.672, df=62.658, P=0.010$)、5 d ($t=3.500, df=78, P=0.001$) 后达到了显著水平,分别降低了 (1.55±0.57) d、(1.83±0.64) d[图2-3(B)]。

在高温类型38℃/10 min下,单独高温持续天数[图2-3(C)灰柱, $F_{2,117}=5.849, P=0.004$]与高温天数和吡虫啉同时作用[图2-3(C)黑柱, $F_{2,117}=5.192, P=0.007$]均对麦长管蚜当代寿命产生显著影响,且均随着高温天数的增加寿命缩短;且在该类型下,无论高温处理持续几天,吡虫啉的存在都延长了麦长管蚜的寿命,且在持续3 d达到了显著水平($t=-2.166, df=78, P=0.033$),相较热单作寿命延长了(2.05±1.13) d[图2-3(C)]。

2.2.2.2 繁殖

高温类型、持续天数和吡虫啉对麦长管蚜繁殖的三因子方差分析结果表明(表2-3):高温类型和持续天数对麦长管蚜当代繁殖影响差异显著。而吡虫啉有无仅对麦长管蚜当代繁殖影响差异不显著。高温类型与持续天数以及高温类型与吡虫啉对麦长管蚜当代繁殖都存在显著的交互作用,但持续天数与吡虫啉以及高温类型、持续天数和吡虫啉对该性状却无交互影响。

表2-3 高温与吡虫啉互作对麦长管蚜当代繁殖的方差分析

Table 2-3 Results of variance analysis for effects of heat events, imidaclopridand their interaction on maternal fecundity of S.avenae

来源 Source	DF	MS	F	P
高温类型 Heat treatment	2	4.34	7.57	0.001
持续天数 Duration	2	8.90	15.50	<0.001
吡虫啉 Imidacoprid	1	0.34	0.59	0.443
高温 × 持续天数 Heat treatment × Duration	4	1.67	2.92	0.021
高温 × 吡虫啉 Heat treatment × Imidacoprid	2	4.16	7.24	0.001
持续天数 × 吡虫啉 Duration × Imidacoprid	2	1.14	1.99	0.137

续表

来源 Source	DF	MS	F	P
高温 × 持续天数 × 吡虫啉 Heat treatment × Duration × Imidacoprid	4	0.35	0.60	0.660
误差 Error	702	0.57	—	—

与空白对照成蚜繁殖力（11.03 ± 1.34）头若蚜/成蚜相比，所有处理均显著压低了当代麦长管蚜的寿命，最少压低了（5.05 ± 1.76）头（图 2-4）。

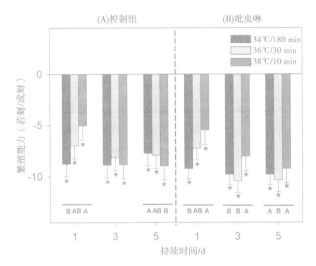

图 2-4　两种农药处理下，3 种高温类型对麦长管蚜当代繁殖的影响（平均值 ± 标准误）。每个柱子表示不同处理组减去空白对照组对应的平均值与标准误。"*"表示处理组和空白对照组间差异达显著水平（$P=0.05$）。大写字母代表不同色柱高温类型间的差异显著性（$P=0.05$）。

Fig. 2-4 Effect of three heat wave types on maternal fecundity (mean ± SE) of S.avenae under pesticide treatment (absent versus present). Each bar represents the mean value and standard error corresponding to the different treatment minus the CK treatment.The "*" represent significant difference ($P=0.05$) between treatment and CK group. Different letters indicate significant level ($P=0.05$) of values between different heat wave types.

无吡虫啉处理的情况下，三种高温类型（34 ℃ /180 min，36 ℃ /30 min，38 ℃ /10 min）在持续 1 d（$F_{2,117} = 3.674, P = 0.028$）、

5 d（$F_{2,117} = 2.885$，$P = 0.048$）后对麦长管蚜当代繁殖影响有显著差异，在持续 3 d（$F_{2,117} = 1.090$，$P = 0.340$）后无差异 [图 2-4(A)]；在持续 1 d 后，34℃/180 min 产生的负面影响最大，产仔量仅为（2.25 ± 0.49）d；在持续 5 d 后，38℃/10 min 产生的负面影响最大，产仔量仅为（2.05 ± 0.66）d。

在吡虫啉吡虫啉存在的条件下，三种高温类型在持续 1 d（$F_{2,117} = 2.620$，$P = 0.045$）、3 d（$F_{2,117} = 7.720$，$P < 0.001$）、5 d（$F_{2,117} = 5.687$，$P = 0.004$）后均对麦长管蚜当代繁殖影响有显著差异，且无论在哪个持续天数下，38℃/10 min 和吡虫啉互作产生的负面影响均最小 [图 2-4(B)]。

在高温类型 34℃/180 min 下，单独高温持续天数 [图 2-5(A) 灰柱，$F_{2,117} = 0.957$，$P = 0.387$] 以及高温天数和吡虫啉同时作用 [图 2-5(A) 黑柱，$F_{2,117} = 1.294$，$P = 0.278$] 均没有对麦长管蚜当代繁殖产生显著影响；且在该类型下，无论热持续几天，吡虫啉的存在都抑制了麦长管蚜当代繁殖，且在持续 3 d（$t = 2.050$，$df = 78$，$P = 0.044$）、5 d（$t = 2.594$，$df = 78$，$P = 0.011$）达到了显著水平，分别降低了（0.90 ± 0.54）头、（2.17 ± 0.91）头 [图 2-5(A)]。

在高温类型 36℃/30 min 下，单独高温持续天数 [图 2-5(B) 灰柱，$F_{2,117} = 0.868$，$P = 0.422$] 对麦长管蚜当代繁殖无显著影响，但高温天数和吡虫啉同时作用 [图 2-5(B) 黑柱，$F_{2,117} = 11.684$，$P < 0.001$] 却对麦长管蚜当代繁殖产生显著影响，高温持续 1 d 和吡虫啉同时作用后，麦长管蚜的繁殖为（3.78 ± 0.89）头，显著高于高温持续 3 d、5 d 和吡虫啉同时作用后麦长管蚜（0.55 ± 0.16）头和（0.62 ± 0.21）头；且在该类型下，无论热持续几天，吡虫啉的存在都抑制了当代麦长管蚜的繁殖，与高温类型 34℃/180 min 下趋势相同，且在持续 3 d（$t = 3.968$，$df = 78$，$P < 0.001$）、5 d（$t = 5.324$，$df = 78$，$P < 0.001$）达到了显著水平，分别降低了（2.33 ± 0.65）头、（2.45 ± 0.61）头 [图 2-5(B)]。

在高温类型 38℃/10 min 下，单独高温持续天数 [图 2-5(C) 灰柱，$F_{2,117} = 5.849$，$P = 0.001$] 与高温天数和吡虫啉同时作用 [图 2-5(C) 黑柱，$F_{2,117} = 5.192$，$P = 0.007$] 均对麦长管蚜当代繁殖产生显著影响；且在该类型下，吡虫啉的有无在高温持续 1 d

($t = 0.562, df = 78, P = 0.576$)、3 d($t = -1.302, df = 78, P = 0.197$)、5 d($t = -0.452, df = 78, P = 0.652$)后对当代繁殖均无影响[图2-5(C)]。

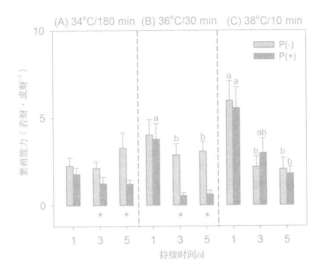

图2-5 3种高温类型处理下,高温持续时间和吡虫啉对麦长管蚜当代繁殖的影响(平均值 ± 标准误)。"*"表示不同色柱吡虫啉处理间差异达显著水平($P=0.05$)。小写字母代表同一色柱高温持续天数间的差异显著性($P=0.05$)。

Fig. 2-5 Maternal fecundity (mean ± SE) of S.avenae at different heat wave duration interaction with imidacloprid. "*" indicate significance level ($P=0.05$) of values between pesticide treatment. Different letters indicate significant level ($P=0.05$) of values between different heat wave duration.

2.2.2.3 繁殖率

高温类型、持续天数和吡虫啉对麦长管蚜当代繁殖率的3个因子方差分析结果表明(表2-4):高温类型对麦长管蚜当代繁殖率没有产生显著影响,但不同高温持续天数和吡虫啉处理却对麦长管蚜当代繁殖率影响显著。高温与持续天数以及持续天数和吡虫啉对麦长管蚜当代繁殖率均不存在显著的交互作用,但高温与吡虫啉以及高温、持续天数和吡虫啉对该性状却存在显著交互作用。

表 2-4　高温与吡虫啉互作对麦长管蚜当代繁殖率的方差分析

Table 2-4 Results of variance analysis for effects of heat events, imidacloprid and their interaction on maternal fecundity rate of S.avenae

来源 Source	DF	MS	F	P
高温类型 Heat treatment	2	0.07	0.40	0.674
持续天数 Duration	2	1.10	5.86	0.003
吡虫啉 Imidacoprid	1	5.25	27.99	<0.001
高温 × 持续天数 Heat treatment × Duration	4	0.36	1.93	0.104
高温 × 吡虫啉 Heat treatment × Imidacoprid	2	1.90	10.15	<0.001
持续天数 × 吡虫啉 Duration × Imidacoprid	2	0.09	0.46	0.631
高温 × 持续天数 × 吡虫啉 Heat treatment × Duration × Imidacoprid	4	0.57	3.03	0.017
误差 Error	702	0.19	—	—

与空白对照繁殖率（2.95±0.15）头若蚜/成蚜/天相比，所有处理均显著压低了当代麦长管蚜的繁殖率，最少压低了每天（0.55±0.24）头若蚜/成蚜（图2-6）。

无吡虫啉处理的情况下，三种高温类型（34℃/180 min，36℃/30 min，38℃/10 min）仅在持续5 d后对麦长管蚜当代繁殖率影响有显著差异（$F_{2,117}$ = 5.008，P = 0.008），且38℃/10 min产生的负面影响最大，麦长管蚜的繁殖率仅为每天（1.71±0.11）头若蚜/成蚜；但在持续1 d（$F_{2,117}$ = 2.010，P = 0.139）、3 d（$F_{2,117}$ = 1.952，P = 0.147）三种高温类型对该性状的影响却无显著差异[图2-6(A)]。

在吡虫啉存在的条件下，三种高温类型（34℃/180 min，36℃/30 min，38℃/10 min）持续1 d后对麦长管蚜当代繁殖率影响无显著差异（$F_{2,117}$ = 0.484，P = 0.617）；但在持续3 d（$F_{2,117}$ = 5.265，P = 0.006）、5 d（$F_{2,117}$ = 7.350，P = 0.001）三种高温类型对该性状的影响却存在显著差异[图2-6(B)]。

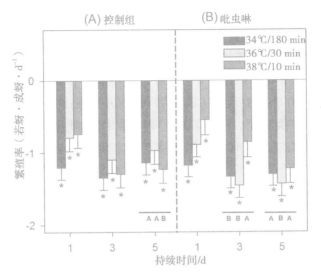

图 2-6 两种农药处理下,3 种高温类型对麦长管蚜当代繁殖率的影响(平均值 ± 标准误)。每个柱子表示不同处理组减去空白对照组对应的平均值与标准误。"*"表示处理组和空白对照组间差异达显著水平(P=0.05)。大写字母代表不同色柱高温类型间的差异显著性(P=0.05)。

Fig. 2-6 Effect of three heat wave types on maternal fecundity rate (mean ± SE) of *S.avenae* under pesticide treatment (absent versus present). Each bar represents the mean value and standard error corresponding to the different treatment minus the CK treatment. The "*" represent significant difference (P=0.05) between treatment and CK group. Different letters indicate significant level (P=0.05) of values between different heat wave types.

在高温类型 34℃/180 min 下,单独高温持续天数[图 2-7(A)灰柱, $F_{2,117}$=1.362,P=0.260] 以及高温天数和吡虫啉同时作用[图 2-7(A)黑柱, $F_{2,117}$=1.503,P=0.227] 均没有对麦长管蚜当代繁殖率产生显著影响;且在该类型下,吡虫啉的有无在高温持续 1 d (t=0.708,df=78,P=0.481)对当代繁殖率均无显著影响,但在高温持续 3 d (t=2.618,df=74.837,P=0.011)、5 d (t=2.473,df=78,P=0.016)后对当代繁殖率产生了显著影响[图 2-7(A)]。

高温类型 36℃/30 min 下,单独高温持续天数[图 2-7(B)灰柱, $F_{2,117}$=0.703,P=0.497] 对麦长管蚜当代繁殖率无显著影响,但高温天数和吡虫啉同时作用 [图 2-7(B) 黑柱,$F_{2,117}=7.509,P=0.001$] 却对麦长管蚜当代繁殖率产生显著影响,高温持续 1 d 和吡虫啉同时作用后,麦长管蚜的繁殖率为每天(0.43±0.07)头若蚜 / 成蚜,显著高于高温持续 3 d、5 d 和吡虫啉同时作用后麦长管蚜每天(0.16±0.05)头若蚜 / 成蚜和每天(0.13±0.04)头若蚜 / 成蚜;且在该类型下,无论高温处理持续几天,吡虫啉的存在相较高温单作均对麦长管蚜的繁殖率产生了负面影响,且在持续 3 d($t=3.797,df=72.482,P<0.001$)、5 d($t=5.920,df=78,P<0.001$)后达到了显著水平,每天分别降低了(0.35±0.09)头若蚜 / 成蚜(0.47±0.08)头若蚜 / 成蚜 [图 2-7(B)]。

在高温类型 38℃/10 min 下,单独高温持续天数 [图 2-7(C) 灰柱,$F_{2,117}=5.480,P=0.005$] 对麦长管蚜当代繁殖率产生显著影响,但高温天数和吡虫啉同时作用 [图 2-7(C) 黑柱,$F_{2,117}=0.098,P=0.907$] 均对麦长管蚜当代繁殖率无显著影响;且在该类型下,吡虫啉的有无在高温持续 1 d($t=1.526, df=78, P=0.131$)、3 d($t=-1.133, df=76.066, P=0.261$)、5 d($t=-1.084, df=78, P=0.282$)后对当代繁殖率均无影响(图 2-7C)。

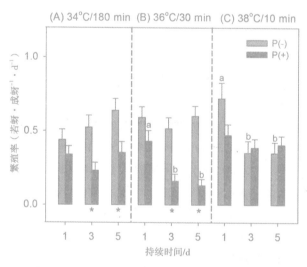

图 2-7 3 种高温类型处理下,高温持续时间和吡虫啉对麦长管蚜当代繁殖率的影响(平均值 ± 标准误)。"*"表示不同色柱吡虫啉处理间差异达显著水平($P=0.05$)。小写字母代表同一色柱高温持续天数间的差异显著性($P=0.05$)

Fig. 2-7 Maternal fecundity rate (mean ± SE) of S.avenae at different heat wave duration interaction with imidacloprid. "*" indicate significance level (P=0.05) of values between pesticide treatment. Different letters indicate significant level (P=0.05) of values between different heat wave duration.

2.2.2.4 种群参数

与空白对照麦长管蚜的种群参数相比,经高温农药处理后平均世代时间(ΔG)显著延长,种群净繁殖率(R_0)既有下降也有增长,种群内禀增长率(r_m)显著下降[图 2-8(A)]。

在无吡虫啉存在的条件下,高温持续 1 d 后,对种群参数产生负面影响最大的高温类型是 34℃/180 min,但当热持续天数增加到 3 d、5 d 后,造成负面影响最大的是 38℃/10 min。在吡虫啉存在的条件下,无论高温持续天数是 1 d、3 d 还是 5 d,38℃/10 min 对当代种群参数造成的负面影响均是最小的(图 2-8A)。

在高温类型 34℃/180 min 下,随着高温持续天数的增加,麦长管蚜的种群内禀增长率 r_m、平均世代周期 G、净增长率 R_0 均出现了不同程度的增加,但随着吡虫啉加入,出现了相反的趋势,随着高温天持续天数的增加,种群参数受到了明显的抑制,在高温类型 36℃/30 min 和 38℃/10 min 条件下,随着热持续天数的增加,麦长管蚜当代种群参数均表现出明显下降的趋势,且随着吡虫啉的加入,这种下降的趋势仍然存在[图 2-8(B)]。

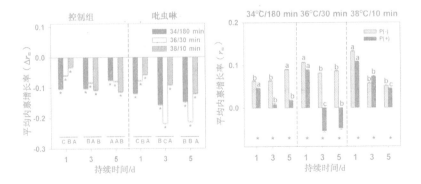

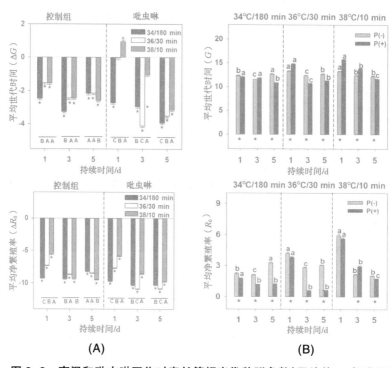

图2-8 高温和吡虫啉互作对麦长管蚜当代种群参数(平均值 ± 标准误)的影响。(A)两种农药处理下,3种高温类型对麦长管蚜当代种群参数的影响。每个柱子表示不同处理组减去空白对照组对应的平均值与标准误。"*"表示处理组和空白对照组间差异达显著水平($P=0.05$)。大写字母代表不同色柱高温类型间的差异显著性($P=0.05$)。(B)3种高温类型处理下,高温持续时间和吡虫啉对麦长管蚜当代种群参数的影响(平均值 ± 标准误)。"*"表示不同色柱吡虫啉处理间差异达显著水平($P=0.05$)。小写字母代表同一色柱高温持续天数间的差异显著性($P=0.05$)。

Fig.2-8 Effects of heat events, imidacloprid and their interaction on maternal population parameters (mean ± SE) of *S.avenae*. (A) Effect of three heat wave types on maternal population parameters of *S.avenae* under pesticide treatment (absent versus present). Each bar represents the mean value and standard error corresponding to the different treatment minus the CK treatment. The "*" represent significant difference ($P=0.05$) between treatment and CK group. Different letters indicate significant level ($P=0.05$) of values between different heat wave types. (B) Maternal population parameters of *S.avenae* at different heat wave duration interaction with imidacloprid. "*" indicate significance level ($P=0.05$) of values between pesticide treatment. Different letters indicate significant level ($P=0.05$) of values between different heat wave duration.

2.2.3 高温与吡虫啉互作对跨代生活史性状的影响

2.2.3.1 子代发育历期

母代经历不同高温类型、不同热持续天数和吡虫啉处理对子代若蚜发育历期的三因子方差分析结果表明(表 2-5)：母代经历单独高温类型、不同热持续天数处理后对子代发育历期产生显著影响，但母代经历单独吡虫啉处理后对子代该性状无显著影响。母代经历高温类型与持续天数以及持续天数与吡虫啉互作对子代若蚜发育历期存在显著的影响，但不存在母代经历高温类型和吡虫啉的交互作用。母代经历不同高温类型、不同持续天数和吡虫啉三因子处理后对子代若蚜发育历期不存在交互。

表 2-5 母代经历高温与吡虫啉互作对子代发育历期的方差分析

Table 2-5 Results of variance analysis for effects of maternal exposing heat events, imidacloprid and their interaction on offspring nymphal time of *S.avenae*

母代处理 Maternal treatment	DF	MS	F	P
高温类型 Heat treatment	2	0.22	5.08	0.07
持续天数 Duration	2	0.57	13.31	<0.001
吡虫啉 Imidacoprid	1	0.10	0.25	0.62
高温 × 持续天数 Heat treatment × Duration	4	0.21	4.85	0.001
高温 × 吡虫啉 Heat treatment × Imidacoprid	2	0.05	1.06	0.35
持续天数 × 吡虫啉 Duration × Imidacoprid	2	0.31	7.26	0.001
高温 × 持续天数 × 吡虫啉 Heat treatment × Duration × Imidacoprid	4	0.04	0.94	0.45
误差 Error	231	0.04	—	—

与空白对照所产子代的发育历期（8.72±0.22）d 相比，母代经高温农药处理后均不同程度的延长了子代发育所需时间（0.05±0.29）~（3.00±0.49 d），除母代经 36℃/30 min 持续 3 d 与吡虫啉互作（$t = -1.072$, $df = 30$, $P = 0.292$）、38℃/10 min 持续 5 d 与吡虫啉互作（$t = -0.179$, $df = 36$, $P = 0.859$）处理外，空白与其余处理相比均达到了显著水平（图 2-9）。

母代在无吡虫啉处理的情况下，经历三种高温类型（34℃/180 min, 36℃/30 min, 38℃/10 min）仅在持续 1 d 后对子代发育历期的影响有显著差异（$F_{2,52} = 4.588$, $P = 0.015$），且母代经历 36℃/30 min 会显著延长子代的发育历期到（9.62±0.35）d，这种趋势在持续 3 d 后（$F_{2,37} = 2.337$, $P = 0.111$）仍然存在但并不显著了，直到 5 d（$F_{2,42} = 0.589$, $P = 0.560$）后这种趋势彻底消失了 [图 2-9(A)]。

母代在吡虫啉存在的条件下，经历三种高温类型在持续 1 d（$F_{2,37} = 1.490$, $P = 0.239$）后对子代发育历期影响无显著差异，但随着持续天数的增加差异达到显著，如在持续 3 d（$F_{2,24} = 3.813$, $P = 0.036$）、5 d（$F_{2,39} = 8.642$, $P = 0.001$）[图 2-9(B)]。

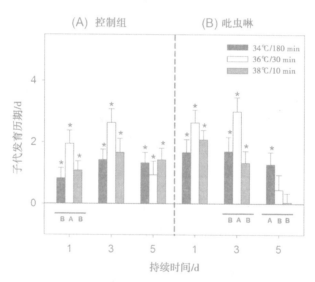

图 2-9 母代在两种农药处理下，经历 3 种高温类型处理后对麦长管蚜子代发育历期的影响（平均值±标准误）。每个柱子表示不同处理组减去空白对照组对应的平均值与标准误。"*" 表示处理组和空白对照组间差异达显著水

平（$P=0.05$）。大写字母代表不同色柱高温类型间的差异显著性（$P=0.05$）。

Fig. 2-9 Effect of maternal exposing three heat wave types on offspring nymphal time (mean ± SE) of *S.avenae* under pesticide treatment (absent versus present). Each bar represents the mean value and standard error corresponding to the different treatment minus the CK treatment. The "*" represent significant difference ($P=0.05$) between treatment and CK group. Different letters indicate significant level ($P=0.05$) of values between different heat wave types.

母代在高温类型 34℃/180 min 处理下，经历不同持续天数 [图 2-10(A) 灰柱，$F_{2,62} = 1.475, P = 0.237$] 以及不同持续天数和吡虫啉同时作用 [图 2-10(A) 黑柱，$F_{2,43} = 0.319, P = 0.728$] 均没有对子代若蚜历期产生影响；且在该类型下，母代是否经吡虫啉处理在高温持续 1 d（$t = -1.505, df = 35, P = 0.141$）、3 d（$t = -0.522, df = 30, P = 0.606$）、5 d（$t = 0.087, df = 40, P = 0.931$）均对子代若蚜历期无影响 [图 2-10(A)]。

母代在高温类型 36℃/30 min 处理下，经历不同持续天数 [图 2-10(B) 灰柱，$F_{2,36} = 4.453, P = 0.019$] 以及不同持续天数和吡虫啉同时作用 [图 2-10(B) 黑柱，$F_{2,27} = 8.994, P = 0.001$] 均对子代若蚜历期产生显著影响；且与母代在高温类型 34℃/180 min 下相同，在该类型处理下，母代是否经吡虫啉处理在高温持续 1 d（$t = -1.330, df = 24, P = 0.196$）、3 d（$t = -0.523, df = 20, P = 0.606$）、5 d（$t = 1.080, df = 19, P = 0.294$）均对子代若蚜历期无影响 [图 2-10(A)]。

母代在高温类型 38℃/10 min 处理下，经历不同持续天数 [图 2-1(C) 灰柱，$F_{2,33} = 0.806, P = 0.455$] 对子代若蚜历期无显著影响，但不同持续天数和吡虫啉同时作用 [图 2-1(C) 黑柱，$F_{2,30} = 18.076, P < 0.001$] 却对子代若蚜历期产生显著影响；且在该类型下，母代是否经吡虫啉处理在高温持续 1 d（$t = -2.955, df = 30, P = 0.006$）、5 d（$t = 3.082, df = 22, P = 0.005$）对子代若蚜历期产生显著影响，但在持续 3 d（$t = 0.0640, df = 11, P = 0.535$）却无显著影响 [图 2-10(C)]。

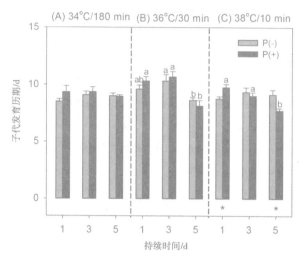

图2-10 母代在3种高温类型处理下,经历不同高温持续时间和吡虫啉互作对麦长管蚜子代发育历期的影响(平均值 ± 标准误)。"*"表示不同色柱吡虫啉处理间差异达显著水平($P=0.05$)。小写字母代表同一色柱高温持续天数间的差异显著性($P=0.05$)。

Fig. 2-10 Offspring nymphal time (mean ± SE) of *S. avenae* when maternal at different heat wave duration interaction with imidacloprid. "*" indicate significance level ($P=0.05$) of values between pesticide treatment. Different letters indicate significant level ($P=0.05$) of values between different heat wave duration.

2.2.3.2 子代产仔前期

母代经历不同高温类型、不同热持续天数和吡虫啉处理对子代产仔前期的三因子方差分析结果表明(表2-6):母代经历单独高温类型、不同热持续天数以及吡虫啉处理后对子代产仔前期均没有产生显著影响。且这三个因素无论是两两交互还是三者交互均不会对子代产仔前期产生影响。且与空白对照所产子代的产仔前期相比,母代经历无论是经历高温、农药单作还是互作处理对子代该性状均无显著影响。

表2-6 母代经历高温与吡虫啉互作对子代产仔前期的方差分析

Table 2-6 Results of variance analysis for effects of maternal exposing heat events, imidacloprid and their interaction on offspring pre-productive period of *S.avenae*

母代处理 Maternal treatment	DF	MS	F	P
高温类型 Heat treatment	2	0.36	1.16	0.32
持续天数 Duration	2	0.06	0.2	0.82
吡虫啉 Imidacoprid	1	0.8	2.56	0.11
高温 × 持续天数 Heat treatment × Duration	4	0.07	0.22	0.93
高温 × 吡虫啉 Heat treatment × Imidacoprid	2	0.24	0.76	0.47
持续天数 × 吡虫啉 Duration × Imidacoprid	2	0.72	2.3	0.10
高温 × 持续天数 × 吡虫啉 Heat treatment × Duration × Imidacoprid	4	0.74	2.38	0.05
误差 Error	222	0.31	—	—

2.2.3.3 子代若虫存活

与空白对照所产子代若虫存活率80%相比,母代经高温、吡虫啉所有处理后均不同程度的压低了子代的若虫存活率,除母代经高温34℃/30 min持续1 d热单作($\chi^2 = 0.373$, $df = 1$, $P = 0.542$)、3 d热单作($\chi^2 = 0.800$, $df = 1$, $P = 0.371$)、5 d($\chi^2 = 0.373$, $df = 1$, $P = 0.542$)热单作,高温38℃/10 min持续1 d热单作($\chi^2 = 1.364$, $df = 1$, $P = 0.243$)以及高温34℃/10 min持续5 d与吡虫啉互作($\chi^2 = 1.364$, $df = 1$, $P = 0.243$)处理外,空白与其余处理相比均达到了显著水平(图2-11)。

母代在无吡虫啉处理的情况下,经历三种高温类型(34℃/180 min,36℃/30 min,38℃/10 min)在持续1 d ($\chi^2 = 6.265$, $df = 2$, $P = 0.044$)、3 d ($\chi^2 = 15.210$, $df = 2$, $P < 0.001$)、5 ($\chi^2 = 10.167$, $df = 2$, $P = 0.006$)后均会对子代若蚜存活产生显著影响。且无论高温持续几天,34℃/180 min高温类型对子代若蚜存活产生的负面影响均最小[图2-11(A)]。

母代在吡虫啉存在的条件下,经历三种高温类型在持续1 d

（$\chi^2 = 0.630$, $df = 2$, $P = 0.730$）、3 d（$\chi^2 = 1.270$, $df = 2$, $P = 0.530$）后对子代若蚜存活均没有产生显著差异，但当持续 5 d 后（$\chi^2 = 9.042$, $df = 2$, $P = 0.011$）却对该性状产生了显著影响；且与母代无吡虫啉处理的情况相同，无论高温持续几天，34℃/180 min 高温类型对子代若蚜存活产生的负面影响均最小 [图 2-11(B)]。

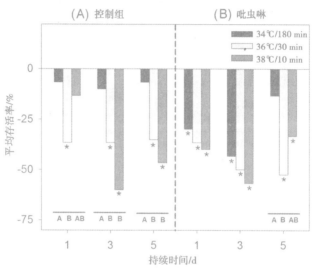

图 2-11 母代在两种农药处理下，经历 3 种高温类型处理后对麦长管蚜子代若虫存活的影响（平均值 ± 标准误）。每个柱子表示不同处理组减去空白对照组对应的平均值与标准误。"*"表示处理组和空白对照组间差异达显著水平（$P=0.05$）。大写字母代表不同色柱高温类型间的差异显著性（$P=0.05$）。

Fig. 2-11 Effect of maternal exposing three heat wave types on offspring survival rate of S.avenae under pesticide treatment (absent versus present). Each bar represents the mean value and standard error corresponding to the different treatment minus the CK treatment. The "*" represent significant difference ($P=0.05$) between treatment and CK group. Different letters indicate significant level ($P=0.05$) of values between different heat wave types.

在高温类型 34℃/180 min 处理下，母代经历单独高温持续天数对子代若蚜存活的影响差异不显著 [图 2-12(A) 灰柱，$\chi^2 = 0.111$, $df = 2$, $P = 0.946$]，但经历高温天数和农药同时作用 [图 2-12(A) 黑柱，$\chi^2 = 5.425$, $df = 2$, $P = 0.066$] 却对该性状产生临界显著影响;且在该类型处理下,母代经历吡虫啉的处理后均会降低子代若蚜的存活率,甚至在热持续 3 d 后达到了显著水平($\chi^2 = 6.696$, $df = 1$, $P = 0.010$),存活率下降了 33.3%[图 2-12(A)]。

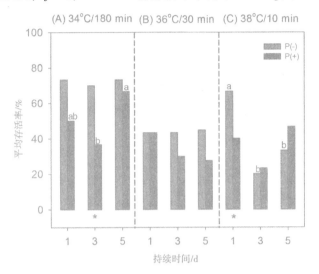

图 2-12 母代在 3 种高温类型处理下,经历不同高温持续时间和吡虫啉互作对麦长管蚜子代若虫存活的影响(平均值 ± 标准误)。"*"表示不同色柱吡虫啉处理间差异达显著水平(P=0.05)。小写字母代表同一色柱高温持续天数间的差异显著性(P=0.05)。

Fig. 2-12 Offspring survival rate of *S. avenae* when maternal at different heat wave duration interaction with imidacloprid. "*" indicate significance level (P=0.05) of values between pesticide treatment. Different letters indicate significant level (P=0.05) of values between different heat wave duration.

在高温类型 36℃/30 min 处理下,母代经历单独高温持续天数 [图 2-12(B) 灰柱, $\chi^2 = 0.018$, $df = 2$, $P = 0.991$] 以及高温天数和农药同时作用 [图 2-12(B) 黑柱, $\chi^2 = 0.055$, $df = 2$, $P = 0.973$] 对子代若蚜存活的影响差异均不显著;且在该类型处理下,母代经历吡虫啉的处理后在热持续天数为 3 d、5 d 后均会降低子代若蚜的存活率,分别为 13.3%、17.2%,但均没有达到显著水平($P > 0.05$)[图 2-12(B)]。

在高温类型 38℃/10 min 处理下,母代经历单独高温持续天数对子代若蚜存活的影响差异显著[图 2-12(C)灰柱，χ^2 = 14.444, df = 2, P = 0.001]，但经历高温天数和农药同时作用后[图 2-12(C)黑柱，χ^2 = 3.732, df = 2, P = 0.155]却对该性状无显著影响；且在该类型处理下，母代经历吡虫啉处理在热持续 1 d 后会显著降低子代若蚜的存活率（χ^2 = 4.286, df = 1, P = 0.038），使得存活率下降了 20.7%，但是热持续 3 d、5 d 后，却出现了反转，母代经历吡虫啉处理反而提高了若蚜的存活率，分别提高了为 3.3%、13.4%，只是未达到显著水平[图 2-12(C)]。

2.2.3.4 子代成蚜寿命

母代经历不同高温类型、不同热持续天数和吡虫啉处理对子代成蚜寿命的三因子方差分析结果表明（表 2-7）：母代经历单独高温类型、不同热持续天数以及吡虫啉处理后对子代成蚜寿命均没有产生显著影响。且这三个因素无论是两两交互还是三者交互均不会对子代产仔前期产生影响。母代经历不同高温类型与持续天数以及高温类型和吡虫啉互作对子代成蚜寿命存在显著的影响，但不存在母代经历高温天数与吡虫啉的交互作用。母代经历不同高温类型、不同持续天数和吡虫啉三因子处理后对子代成蚜寿命不存在交互。

表 2-7 母代经历高温与吡虫啉互作对子代成蚜寿命的方差分析

Table 2-7 Results of variance analysis for effects of maternal exposing heat events, imidacloprid and their interaction on offspring longevity of *S. avenae*

母代处理 Maternal treatment	DF	MS	F	P
高温类型 Heat treatment	2	98.55	2.80	0.06
持续天数 Duration	2	47.32	1.35	0.26
吡虫啉 Imidacoprid	1	60.72	1.73	0.19
高温 × 持续天数 Heat treatment × Duration	4	85.58	2.44	0.05
高温 × 吡虫啉 Heat treatment × Imidacoprid	2	121.86	3.47	0.03
持续天数 × 吡虫啉 Duration × Imidacoprid	2	78.57	2.24	0.11
高温 × 持续天数 × 吡虫啉 Heat treatment × Duration × Imidacoprid	4	57.74	1.64	0.16
误差 Error	231	35.15	—	—

与空白对照所产子代寿命相比,所有处理均对子代的寿命产生了刺激作用,且在母代经 36℃/30 min 持续 5 d 热单作（ $t = -4.279$, $df = 35$, $P<0.001$ ）、34℃/180 min 持续 1 d（ $t = -4.995$, $df = 36.477$, $P<0.001$ ）、3 d（ $t = -3.042$, $df = 33$, $P=0.005$ ）、5 d（ $t = -3.477$, $df = 42$, $P=0.001$ ）与吡虫啉互作以及 36℃/30 min 持续 3 d 与吡虫啉互作（ $t = -5.035$, $df = 31$, $P<0.001$ ）处理下均达到了显著水平（图 2-13）。

母代在无吡虫啉处理的情况下,经历三种高温类型（34℃/180 min, 36℃/30 min, 38℃/10 min）在持续 1 d（ $F_{2,52} = 0.278$, $P = 0.758$ ）、3 d（ $F_{2,42} = 0.883$, $P = 0.422$ ）后对子代成蚜寿命都没有显著影响,仅在持续 5 d 后对该性状产生临界显著影响（ $F_{2,42} = 2.833$, $P = 0.070$ ）[图 2-13(A)]。

母代在吡虫啉存在的条件下,与未经吡虫啉处理的情况刚好相反,经历三种高温类型在持续 1 d（ $F_{2,37} = 9.076$, $P = 0.001$ ）、3 d（ $F_{2,24} = 4.614$, $P = 0.020$ ）后对子代寿命产生显著影响,但随着持续天数的增加到 5 d 后（ $F_{2,39} = 2.232$, $P = 0.121$ ）,这种母代高温类型所造成的显著差异消失了 [图 2-13(B)]。

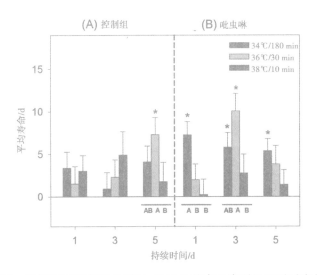

图 2-13 母代在两种农药处理下,经历 3 种高温类型处理后对麦长管蚜子代寿命的影响（平均值 ± 标准误）。每个柱子表示不同处理组减去空白对照组对应的平均值与标准误。"*"表示处理组和空白对照组间差异达显著水平（ $P=0.05$ ）。大写字母代表不同色柱高温类型间的差异显著性（ $P=0.05$ ）。

Fig. 2-13 Effect of maternal exposing three heat wave types on offspring longevity (mean ± SE) of S.avenae under pesticide treatment (absent versus present). Each bar represents the mean value and standard error corresponding to the different treatment minus the CK treatment. The "*" represent significant difference ($P=0.05$) between treatment and CK group. Different letters indicate significant level ($P=0.05$) of values between different heat wave types.

母代在高温类型 34℃/180 min 下，经历不同持续天数[图 2-14(A) 灰柱，$F_{2,62} = 1.285, P = 0.284$]以及不同持续天数和吡虫啉同时作用[图 2-14(A) 黑柱，$F_{2,43} = 0.782, P = 0.464$]均没有对子代成蚜寿命产生显著影响；且在该类型下，母代在所有热持续天数下，相较未经农药处理后的子代成蚜寿命，经吡虫啉处理后均延长了子代成蚜的寿命，甚至在高温持续 3 d（$t = -2.234, df = 30, P = 0.033$）达到了显著水平，延长了（4.87 ± 2.18）d[图 2-14(A)]。

母代在高温类型 36℃/30 min 下，经历不同持续天数[图 2-14(B) 灰柱，$F_{2,36} = 3.172, P = 0.048$]以及不同持续天数和吡虫啉同时作用[图 2-14(B) 黑柱，$F_{2,27} = 6.191, P = 0.006$]均对子代成蚜寿命产生显著影响；且在该类型下，母代是否经吡虫啉处理在高温持续 1 d（$t = -1.930, df = 24, P = 0.848$）、5 d（$t = 1.558, df = 19, P = 0.136$）均对子代成蚜寿命无显著影响，仅在持续 3 d（$t = -2.713, df = 20, P = 0.013$）后产生了显著影响[图 2-14(B)]。

母代在高温类型 38℃/10 min 下，经历不同持续天数[图 2-14(C) 灰柱，$F_{2,62} = 1.285, P = 0.284$]以及不同持续天数和吡虫啉同时作用[图 2-14(C) 黑柱，$F_{2,43} = 0.782, P = 0.464$]均没有对子代成蚜寿命产生显著影响；且在该类型下，母代在所有热持续天数下，相较高温单作处理后的子代成蚜寿命，经吡虫啉处理后均缩短了子代成蚜的寿命，但是这种缩短的幅度随着高温持续天数的增加不断减少，如在高温持续 1 d、3 d、5 d 分别缩短了（2.78 ± 2.20）d、（2.12 ± 2.66）d、（0.34 ± 2.57）d[图 2-14(C)]。

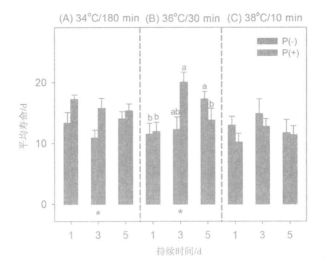

图2-14 母代在3种高温类型处理下,经历不同高温持续时间和吡虫啉互作对麦长管蚜子代寿命的影响(平均值 ± 标准误)。"*"表示不同色柱吡虫啉处理间差异达显著水平($P=0.05$)。小写字母代表同一色柱高温持续天数间的差异显著性($P=0.05$)。

Fig. 2-14 Offspring longevity (mean ± SE) of *S.avenae* when maternal at different heat wave duration interaction with imidacloprid. "*" indicate significance level ($P=0.05$) of values between pesticide treatment. Different letters indicate significant level ($P=0.05$) of values between different heat wave duration.

2.2.3.5 子代成蚜繁殖

母代经历不同高温类型、不同热持续天数和吡虫啉处理对子代成蚜繁殖的三因子方差分析结果表明(表2-8):母代经历单独高温类型以及吡虫啉处理后对子代成蚜繁殖均没有产生显著影响,但经历不同热持续天数却对子代成蚜繁殖产生显著影响。此外,这三个因素无论是两两交互还是三者交互均不会对子代成蚜繁殖产生显著影响。

表 2-8 母代经历高温与吡虫啉互作对子代成蚜繁殖的方差分析

Table 2-8 Results of variance analysis for effects of maternal exposing heat events, imidacloprid and their interaction on offspring fecundity of S.avenae

母代处理 Maternal treatment	DF	MS	F	P
高温类型 Heat treatment	2	672.74	2.58	0.08
持续天数 Duration	2	956.76	3.68	0.03
吡虫啉 Imidacoprid	1	401.39	1.54	0.22
高温 × 持续天数 Heat treatment × Duration	4	315.68	1.21	0.31
高温 × 吡虫啉 Heat treatment × Imidacoprid	2	254.46	0.98	0.38
持续天数 × 吡虫啉 Duration × Imidacoprid	2	524.84	2.02	0.14
高温 × 持续天数 × 吡虫啉 Heat treatment × Duration × Imidacoprid	4	229.51	0.88	0.48
误差 Error	231	260.33	—	—

与空白对照所产子代的繁殖力(21.17±2.71)头若蚜/成蚜相比，除母代经36℃/30 min 和38℃/10 min 持续3 d 热单作、36℃/30 min 和38℃/10 min 持续1 d 和农药互作对子代繁殖有负面影响外，其余处理均对子代的繁殖产生了刺激作用，且在母代经36℃/30 min 持续5 d 热单作($t = -3.215$, $df = 31$, $P = 0.003$)、34℃/180 min 持续1 d 与吡虫啉互作($t = -3.215$, $df = 31$, $P = 0.003$)、34℃/180 min 持续5 d 与吡虫啉互作($t = -3.215$, $df = 31$, $P = 0.003$)处理下达显著水平(图2-15)。

母代在无吡虫啉处理的情况下，经历三种高温类型(34℃/180 min, 36℃/30 min, 38℃/10 min)在持续1 d ($F_{2,52} = 0.575$, $P = 0.566$)、3 d ($F_{2,37} = 0.727$, $P = 0.490$)、5 d ($F_{2,42} = 0.790$, $P = 0.461$)后对子代成蚜繁殖都没有显著影响[图2-15(A)]。

母代在吡虫啉存在的条件下，经历三种高温类型(34℃/180 min, 36℃/30 min, 38℃/10 min)在持续1 d ($F_{2,37} = 4.105$, $P = 0.025$)、5 d ($F_{2,39} = 3.681$, $P = 0.34$)后对子代成蚜繁殖都产生显著影响；但在持续3 d ($F_{2,24} = 0.327$, $P = 0.724$)后对该性状无显著性影响[图2-15(B)]。

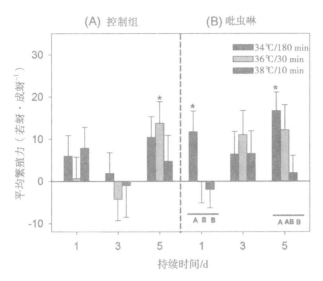

图 2-15 母代在两种农药处理下,经历 3 种高温类型处理后对麦长管蚜子代繁殖的影响(平均值 ± 标准误)。每个柱子表示不同处理组减去空白对照组对应的平均值与标准误。"*"表示处理组和空白对照组间差异达显著水平($P=0.05$)。大写字母代表不同色柱高温类型间的差异显著性($P=0.05$)。

Fig. 2-15 Effect of maternal exposing three heat wave types on offspring fecundity(mean ± SE)of S.avenae under pesticide treatment(absent versus present). Each bar represents the mean value and standard error corresponding to the different treatment minus the CK treatment. The "*" represent significant difference($P=0.05$)between treatment and CK group. Different letters indicate significant level($P=0.05$)of values between different heat wave types.

母代在高温类型 34 ℃ /180 min 下,经历不同持续天数 [图 2-16(A) 灰柱,$F_{2,62} = 1.270$,$P = 0.288$] 以及不同持续天数和吡虫啉同时作用 [图 2-16(A) 黑柱,$F_{2,43} = 1.580$,$P = 0.218$] 均没有对子代成蚜繁殖产生显著影响;且在该类型下,相较无吡虫啉处理单作,母代经吡虫啉处理后无论在高温持续 1 d、3 d、5 d 后尽管未达显著水平($P> 0.05$),但均轻微地刺激子代的繁殖,分别增加了($5.75 ± 5.84$)头、($4.59 ± 5.50$)头、($6.26 ± 5.41$)头 [图 2-16(A)]。

母代在高温类型 36 ℃ /30 min 下,经历不同持续天数 [图

2-16(B)灰柱，$F_{2,36}$ = 4.549, P = 0.017]对子代成蚜繁殖产生显著影响，但不同持续天数和吡虫啉同时作用[图2-16(B)黑柱，$F_{2,27}$ = 2.089, P = 0.143]却对子代成蚜繁殖无显著影响；且在该类型下，母代是否经吡虫啉处理在高温持续 1 d (t = 0.122, df = 24, P = 0.904)、5 d (t = 0.222, df = 19, P = 0.827)对子代成蚜繁殖无显著影响，但在持续 3 d (t = −2.326, df = 20, P = 0.031)却产生显著影响[图2-16(B)]。

母代在高温类型 38℃/10 min 下，与在高温处理 34℃/180 min 下情况相同，母代经历不同持续天数[图2-16(C)灰柱，$F_{2,33}$ = 0.558, P = 0.578]以及不同持续天数和吡虫啉同时作用[图2-16(C)黑柱，$F_{2,30}$ = 1.151, P = 0.330]均没有对子代成蚜繁殖产生显著影响；且在该类型下，母代是否经吡虫啉处理在高温持续 1 d (t = 1.704, df = 29.591, P = 0.099)、3 d (t = −1.348, df = 11, P = 0.205)、5 d (t = 0.402, df = 12.946, P = 0.694)均对子代成蚜繁殖无显著影响[图2-16(C)]。

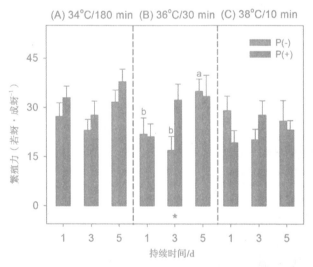

图 2-16 母代在 3 种高温类型处理下，经历不同高温持续时间和吡虫啉对麦长管蚜子代繁殖的影响(平均值 ± 标准误)。"*"表示不同色柱吡虫啉处理间差异达显著水平(P=0.05)。小写字母代表同一色柱高温持续天数间的差异显著性(P=0.05)。

Fig. 2-16 Offspring fecundity (mean ± SE) of *S.avenae* when maternal at different heat wave duration interaction with imidacloprid. "*" indicate significance level (P=0.05) of values between pesticide treatment. Different letters indicate significant level (P=0.05) of values between different heat wave duration.

2.2.3.6 子代成蚜繁殖率

母代经历不同高温类型、不同热持续天数和吡虫啉处理对子代成蚜繁殖的三因子方差分析结果表明(表 2-9)：母代经历单独高温类型以及吡虫啉处理后对子代成蚜繁殖率均没有产生显著影响,但经历不同热持续天数却对子代成蚜繁殖率产生显著影响。此外,这三个因素无论是两两交互还是三者交互均不会对子代成蚜繁殖产生显著影响。

表 2-9 母代经历高温与吡虫啉互作对子代成蚜繁殖率的方差分析

Table 2-9 Results of variance analysis for effects of maternal exposing heat events, imidacloprid and their interaction on offspring fecundity rate of *S.avenae*

母代处理 Maternal treatment	DF	MS	F	P
高温类型 Heat treatment	2	1.07	1.72	0.18
持续天数 Duration	2	3.42	5.46	0.01
吡虫啉 Imidacoprid	1	1.29	2.06	0.15
高温 × 持续天数 Heat treatment × Duration	4	0.54	0.87	0.48
高温 × 吡虫啉 Heat treatment × Imidacoprid	2	0.05	0.08	0.92
持续天数 × 吡虫啉 Duration × Imidacoprid	2	0.42	0.67	0.51
高温 × 持续天数 × 吡虫啉 Heat treatment × Duration × Imidacoprid	4	0.37	0.59	0.67
误差 Error	231	0.63	—	—

与空白对照所产子代的繁殖率为每天(2.09 ± 0.15)头若蚜/成蚜相比,除母代经 34℃/180 min 持续 5 d 热单作、38℃/10 min 持续 3 d 和农药互作、34℃/180 min 持续 5 d 和农药互作以及 36℃/30 min 持续 5 d 和农药互作对子代繁殖率有微弱正面刺激外,其余处理均对子代的繁殖产生了负面影响,且在母代经 36℃/30 min 持续 3 d 热单作($t = 3.160$, $df = 35$, $P = 0.003$)处理下达显著水平(图 2-17)。

无论母代是否经吡虫啉处理,经历三种高温类型(34℃/180 min, 36℃/30 min, 38℃/10 min)在持续 1 d[图 2-17(A): $F_{2,52} = 1.051$, $P = 0.357$;图 2-17(B): $F_{2,37} = 0.256$, $P = 0.775$)]、3 d[(2-17(A): $F_{2,37} = 1.800$, $P = 0.179$;图 2-17(B): $F_{2,24} = 1.292$, $P = 0.293$)]、5 d[(图 2-17(A): $F_{2,42} = 0.120$, $P = 0.887$;图 2-17(B) $F_{2,39} = 1.150$, $P = 0.327$)]后均对子代成蚜繁殖率没有显著影响(图 2-17)。

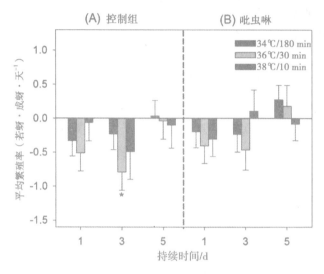

图 2-17 母代在两种农药处理下,经历 3 种高温类型处理后对麦长管蚜子代繁殖率的影响(平均值 ± 标准误)。每个柱子表示不同处理组减去空白对照组对应的平均值与标准误。"*"表示处理组和空白对照组间差异达显著水平($P=0.05$)。大写字母代表不同色柱高温类型间的差异显著性($P=0.05$)。

Fig.2-17 Effect of maternal exposing three heat wave types on offspring fecundityrate(mean ± SE) of S.avenae under pesticide treatment (absent versus present). Each bar represents the mean value and standard error corresponding to the different treatment minus the CK treatment.The "*" represent significant difference ($P=0.05$) between treatment and CK group. Different letters indicate significant level ($P=0.05$) of values between different heat wave types.

母代在高温类型 34℃/180 min 下,经历不同持续天数[图 2-18(A)灰柱,$F_{2,62} = 1.329$,$P = 0.272$]以及不同持续天数和吡虫啉同时作用[图 2-18(A)黑柱,$F_{2,43} = 2.820$,$P = 0.071$]均没

有对子代成蚜繁殖率产生影响;且在该类型下,相较无吡虫啉处理对照,母代经吡虫啉处理后无论在高温持续 1 d、3 d、5 d 后尽管未达显著水平($P > 0.254$),但均轻微地刺激了子代的繁殖率[图 2-18(A)]。

母代在高温类型 36 ℃/30 min 下,经历不同持续天数[图 2-18(B)灰柱,$F_{2,36} = 2.947$, $P = 0.065$]对子代成蚜繁殖率产生临界显著影响,但不同持续天数和吡虫啉同时作用[图 2-18(B)黑柱,$F_{2,27} = 1.804$, $P = 0.184$]却对子代成蚜繁殖率无显著影响;且在该类型下,与母代经高温 34 ℃/180 min 处理下相同,相较该高温下单作,母代经吡虫啉处理后无论在高温持续 1 d、3 d、5 d 后尽管未达显著水平($P > 0.05$),但均对子代繁殖率产生了刺激作用,每天分别增加了(0.12 ± 0.36)、(0.33 ± 0.29)、(0.22 ± 0.34)头若蚜/成蚜[图 2-18(B)]。

母代在高温类型 38 ℃/10 min 下,与在高温处理 34 ℃/180 min 下情况相同,母代经历不同持续天数[图 2-18(C)灰柱,$F_{2,33} = 0.442$, $P = 0.647$]以及不同持续天数和吡虫啉同时作用[图 2-18(C)黑柱,$F_{2,30} = 0.802$, $P = 0.458$]均没有对子代成蚜繁殖率产生影响;且在该类型下,母代是否经吡虫啉处理在高温持续 1 d ($t = 0.756$, $df = 30$, $P = 0.456$)、3 d ($t = -1.120$, $df = 11$, $P = 0.286$)、5 d ($t = -0.053$, $df = 22$, $P = 0.959$)后均对子代成蚜繁殖率无影响[图 2-18(C)]。

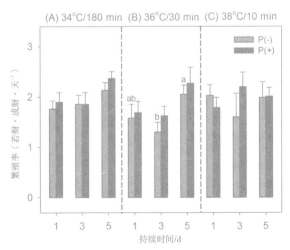

图 2-18 母代在 3 种高温类型处理下,经历不同高温持续时间和吡虫啉对

麦长管蚜子代繁殖率的影响（平均值 ± 标准误）。"*"表示不同色柱吡虫啉处理间差异达显著水平（$P=0.05$）。小写字母代表同一色柱高温持续天数间的差异显著性（$P=0.05$）。

Fig. 2-18 Offspring fecundity rate（mean ± SE）of *S.avenae* when maternal at different heat wave duration interaction with imidacloprid. "*" indicate significance level（$P=0.05$）of values between pesticide treatment. Different letters indicate significant level（$P=0.05$）of values between different heat wave duration.

2.2.3.7 子代种群参数

与空白对照所产子代的种群参数相比，母代经高温农药单作以及互作处理后子代的平均世代时间（ΔG）显著缩短，种群净繁殖率（R_o）显著下降，种群内禀增长率（r_m）显著下降[图 2-19(A)]。

不管母代是否经吡虫啉处理，无论经历高温持续 1 d、3 d 还是 5 d，母代经 34℃/180 min 的高温类型处理后所产子代的平均种群内禀增长率（Δr_m）以及总繁殖率（ΔR_o）相较其余两高温类型处理均增长较快[图 2-19(A)]。

母代经三种高温类型处理后，无论吡虫啉是否存在，在高温持续 3 d 后，子代种群参数受到的负面影响最大，而当高温持续 5 d 后，受到的负面影响最弱[图 2-19(B)]。

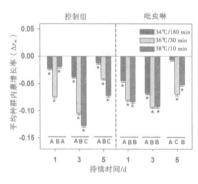
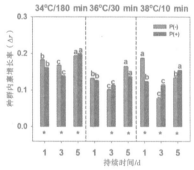

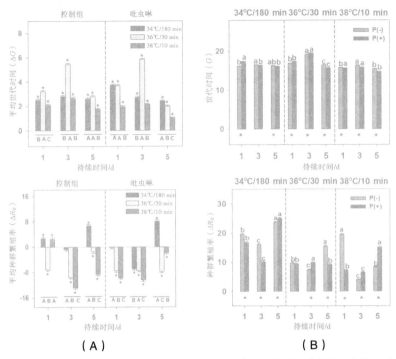

图2-19 母代经历高温和吡虫啉互作对麦长管蚜子代种群参数（平均值 ± 标准误）的影响。（A）母代在两种农药处理下，经历3种高温类型对麦长管蚜子代种群参数的影响。每个柱子表示不同处理组减去空白对照组对应的平均值与标准误。"*"表示处理组和空白对照组间差异达显著水平（$P=0.05$）。大写字母代表不同色柱高温类型间的差异显著性（$P=0.05$）。（B）母代在3种高温类型处理下，经历不同高温持续时间和吡虫啉对麦长管蚜子代种群参数的影响（平均值 ± 标准误）。"*"表示不同色柱吡虫啉处理间差异达显著水平（$P=0.05$）。小写字母代表同一色柱高温持续天数间的差异显著性（$P=0.05$）。

Fig.2-19 Effects of maternal exposingheat events, imidaclopridand their interactionon on offspring population parameters (mean ± SE) of *S.avenae*. (A) Effect of maternal exposing three heat wave types on offspring population parameters of *S.avenae* under pesticide treatment (absent versus present). Each bar represents the mean value and standard error corresponding to the different treatment minus the CK treatment. The "*" represent significant difference ($P=0.05$) between treatment and CK group. Different letters indicate significant level (*P*=0.05) of values between different heat wave types. (B) Offspring population parameters of *S.avenae* at different heat wave duration interaction with imidacloprid. "*" indicate significance level (*P*=0.05) of values between pesticide treatment. Different letters indicate significant level (*P*=0.05) of values between different heat wave duration.

◆ 2.3 结论与讨论

2.3.1 经历高温和吡虫啉双重胁迫后的当代生活史性状响应

高温和吡虫啉对麦长管蚜即时死亡率联合作用评价结果表明：高温类型 34℃/180 min 和 36℃/30 min 较低强度高温、行为反应时间组合与农药互作后对麦长管蚜即时死亡率的影响多数表现为联合毒力的提高（协同、相加），而在高强度高温组合 38℃/10 min 下表现为拮抗作用，持续 1 d、3 d 与吡虫啉互作后对麦长管蚜死亡率的毒力下降。表明这种短时的高强度高温组合可能产生了低剂量兴奋效应，诱导昆虫产生了诸如热激蛋白等应激响应，从而产生了交叉保护作用[163,164]，但随着热持续天数增加到 5 d，胁迫的程度增加，这种保护作用逐渐消失[114,165]。

高温和农药互作除了产生即时效应外，也会影响麦长管蚜后期的生活史性状。相较空白对照而言，经高温、农药胁迫处理后，当代的寿命、繁殖、繁殖率以及种群参数均产生显著的负面响应。但不同高温强度、行为反应时间组合对当代麦长管蚜适合度性状产生了更为复杂的影响。经历 1 d 的热处理后，持续较长时间的温和处理 34℃/180 min 对寿命、繁殖、繁殖率以及内禀增长率产生的负面影响最大，但随着持续时间的增加，38℃/10 min 产生的负面影响最大。这说明，热持续时间所产生的负面影响大于高温强度所造成的影响。但是当吡虫啉同时施用时，持续较长时间的温和（34℃/180 min）或中等（36℃/30 min）高温会比短时极端高温（38℃/10 min）对当代麦长管蚜适合度性状产生的负面影响更显著。

高温持续天数的不同也会对当代麦长管蚜寿命、繁殖、繁殖率以及种群内禀增长率产生复杂的影响。除温和高温组 34℃/180 min,其余两高温类型,无论吡虫啉是否存在,随着热持续天数的增加寿命和繁殖、繁殖率以及种群内禀增长率均会显著下降。对于寿命而言,在高温 34℃/180 min,无论吡虫啉是否存在,热持续天数的增加对寿命均无显著影响,这说明该类型的高温在持续 1 d 后产生的胁迫已使得麦长管蚜的寿命降到了极限,随着持续天数的增加,已没有了降的空间,否则只能死亡,而在该高温类型下,随着持续天数的增加死亡率也会增加也充分说明了这一点。对于繁殖、繁殖率和种群内禀增长率,在高温 34℃/180 min 无吡虫啉存在的情况下,随着热天的增加会出现轻微的增加;但在吡虫啉存在的情况下,繁殖、繁殖率和种群内禀增长率就会随着热天的增加而下降,这可能与两种胁迫本身的性质差异以及昆虫不同的响应机制有关。

农药的有无在不同高温类型下,同样会对当代麦长管蚜寿命、繁殖以及繁殖率产生复杂的影响,但无论在哪种高温类型下,农药的存在均会对种群内禀增长率产生显著负面影响。在高温 34℃/180 min,随着持续天数的增加,吡虫啉的存在都不会对麦长管蚜寿命产生影响,但却会压低繁殖量,说明相较于存活,繁殖对农药的响应更加敏感,对存活影响不大的亚致死农药处理对繁殖也会产生显著影响[86],从而导致繁殖率显著降低。在高温 36℃/30 min,随着持续天数的增加,吡虫啉的存在会显著缩短寿命、压低繁殖,从而导致种群的灭亡,这可能由于在该高温类型下,吡虫啉的毒性得到了提升,加重了对麦长管蚜的负影响,产生了互作效应,符合即时死亡率所得到的联合作用评价。在高温 38℃/10 min,吡虫啉的存在相较无农药处理,对麦长管蚜寿命、繁殖、种群内禀增长率在热持续 3 d 后产生了正面刺激作用,但当持续天数增加到 5 d,农药的这种刺激作用消失,甚至对种群内禀增长率产生负面影响,与即时效应类似原因可能与低剂量兴奋效应有关,但随着胁迫的程度增加,这种保护作用逐渐消失。同样符合即死效应所得到的联合作用评价——两胁迫互作先产生拮抗作用后为相加。

总的来说,仅高温单作时,经历短期 1 d 的不同类型的热处理后,持续较长时间的温和处理 34℃/180 min 对当代生活史性状产生的负面影响最大,但随着持续天数的增加,短时极端高温 38℃/10 min 产生的负面影响最大。但是,当农药存在时,会与温和(34℃/180 min)和中等(36℃/30 min)高温产生显著的互作效应,加重对当代适合度的负面影响;而和极端高温会产生拮抗效应,降低对当代适合度的负面影响。除温和高温组 34℃/180 min,其余两高温类型,无论吡虫啉是否存在,均随着热持续天数的增加,对当代生活史性状产生的负面影响增加。

2.3.2 经历高温和吡虫啉双重胁迫后的跨代生活史性状响应

2.3.2.1 对子代若蚜发育和存活的影响

相较空白对照所产子代,母代无论经历高温单作还是高温农药互作均会延长子代发育所需时间、压低子代若虫存活率。不管吡虫啉是否存在,母代经不同高温类型处理后会对子代若蚜发育、存活产生复杂的影响。吡虫啉不存在的情况下,经历短期 1 d 的不同类型的热处理后,母代经 36℃/30 min 处理对若蚜发育、存活产生的负面影响最大,但随着持续时间的增加,母代经 38℃/10 min 产生的负面影响最大,这更加说明了热对生物的影响不仅取决于热的强度,同样取决于热持续的时间。但是当吡虫啉存在的时候,经历 1 d 和 3 d 的热持续高温后,母代经 36℃/30 min 对若蚜的发育产生的负面影响最大,38℃/10 min 对若蚜存活产生的负面影响最大,热持续经历 5 d 后,母代经 34℃/180 min 对若蚜发育产生的负面影响最大,但 36℃/30 min 对若蚜的存活产生的负面影响最大。这说明,农药和高温两种胁迫同时存在时,相较单独热处理对子代的影响更加复杂。

母代经历不同热持续天数会对子代麦长管蚜若蚜发育、存活产生影响。从子代若蚜发育历期以及存活的结果可知:母代经历不同热持续天数总会对子代这两者其中的一个产生显著影响(除高温 34℃/180 min,无吡虫啉存在处理下),若对发育历期产生显著影响,则对存活的影响不显著,反之亦然;且母代经热持续 3 d,相较热持续 1 d 和 5 d 会延长若蚜发育,降低存活也即随着高温持续天数的增加,对子代发育或存活的负面影响表现先加重后减缓的趋势,这主要是由于随着热持续天数的增加,对母代的损伤逐渐增大,汰选作用逐渐增强,造成母代即时死亡率逐渐增大,但考虑到热持续 1 d 后造成的损伤较小,同时热持续 5 d 后较强的汰选作用,致使存活下的母代个体适合度相差不大,但持续 3 d 这一中等热环境可能会存在适合度较差的母代个体,因此相较热持续 1 d 和 5 d 会对若蚜的发育、存活产生负面影响较强。例如,母代在温和高温 34 ℃ /180 min,无吡虫啉存在处理下,经历热持续 1 d、3 d、5 d 后,子代若蚜的发育历期分别为(8.50 ± 1.06)d、(9.10 ± 1.41)d、(9.00 ± 1.20)d,存活率分别为 73.3%、70.0%、73.3%;而在吡虫啉存在时,母代经历热持续 1 d、3 d、5 d 后,子代若蚜的发育历期分别为(9.33 ± 2.13)d、(9.36 ± 1.43)d、(8.95 ± 0.69)d,存活率分别为 50.0%、36.7%、66.7%。

不同高温类型下,母代是否经农药处理会对子代若蚜发育、存活产生复杂的影响母代在温和(34 ℃ /180 min)和中等(36 ℃ /30 min)高温,经农药处理后相较无农药处理组,均未对子代发育产生影响,但却降低了子代的存活率。而母代在短时极端高温 38 ℃ /10 min,经农药处理后相较无农药处理组,随着热持续天数的增加,对子代的发育和存活的影响趋势相同,对发育从延迟到刺激,对存活从压低到提升,具体而言在高温 38 ℃ /10 min,母代农药处理在热持续 1 d 后,子代发育从(8.75 ± 0.91)d 延长至(9.75 ± 0.97)d,但随着母代经历持续天数增加到 3 d 和 5 d 后,子代发育历期却缩短了,缩短的幅度从(0.33 ± 0.51)d 到(1.39 ± 0.45)d;同样地,该高温类型下,农药的存在在热持续 1 d 后,存活率显著下降,从 66.7% 降至 40.0%,但随着持续天数增加到 3 d 和 5 d 后,农药的存在使得存活率的幅度也随之增加,从 3.3% 升至 13.3%,表明农药与较长时间的极端高温互作后会刺激后代加速发育且产生一定的环境耐受性,提高存活率。

总之,母代经历高温农药胁迫后,除了显著降低母代本身的适合度性状外,这种负面影响也会延续到子代的若虫期发育和存活。母代仅经历高温单作,经历短期 1 d 的不同类型的热处理后,中等高温处理(36 ℃ /30 min)对子代发育、存活所产生的负面影响最大,但随着持续天数的增加,短时极端高温 38 ℃ /10 min 产生的负面影响最大。但是,母代同时经历高温农药胁迫后,高温持续天数较短时(1~3 d),中等和极端高温分别对子代的发育、存活负面影响最大;当高温持续天数达到5 d时,温和和中等高温分别对子代的发育、存活负面影响最大。无论母代是否经农药处理,随着高温持续天数的增加,由于对母代汰选作用逐渐增强的缘故导致对子代发育、存活的负面影响均表现为先加重后缓减的趋势。

2.3.2.2 对子代成蚜表现及种群参数的滞后效应

与若蚜发育历期不同,不管母代经历何种高温类型、何种持续热天数以及是否经历农药均不会对成蚜产仔前期产生影响。

相较空白对照所产子代,母代无论经历高温单作还是高温农药互作,尽管会对子代若虫期(发育和存活)产生负面影响,但这种负面影响却并没有延续到成蚜的生活史性状,甚至会对成蚜性状产生正面刺激作用,如显著延长子代成蚜的寿命,产生这一结果的原因有很多,如补偿效应,是指生物经历胁迫环境后,被转移到适宜环境通过快速的补偿生长得到恢复,中和之前胁迫所带来的负面影响[166,167];模块效应,可以将早期不利的生活环境隔绝,避免对之后的生活阶段产生影响[168,169];毒物兴奋效应,生物经历胁迫后,会对生物后续的相关性状产生正面刺激作用[170,171]。

从高温类型对子代麦长管蚜的成蚜表现的影响可以看出,母代经不同高温类型处理后会对子代成蚜寿命、繁殖产生显著影响但却对繁殖率无显著影响。最重要的是,在与农药互作条件下,高温类型对母代和子代寿命、繁殖的影响存在相反的趋势,当持续1 d高温时,随着高温强度的增加对母代寿命、繁殖的负面影响减弱,出现增加的趋势,而对子代寿命、繁殖的刺激作用下降,表现为逐渐下降的趋势,当持续3 d高温时,中等高温下的母代寿命最短、繁殖量最低,而子代寿命却是最长、繁殖量最高的。这可能是在严酷的环境条件下,母亲可通过缩短自身的寿命、繁殖,将更多的资源分配给子代所致。

除极端高温组处理外(38℃/10 min),无论吡虫啉是否存在,随着母代经历高温持续天数的增加,子代成蚜寿命、繁殖、繁殖率均逐渐提升,且这种现象在36℃中等高温度组中表现更为明显。

母代经高温农药同时处理后,相较母代高温单作,对子代成蚜的寿命、繁殖、繁殖率的影响大部分均不显著,即使显著,经农药处理后也是产生的正面刺激作用。这表明,高温和农药互作后,会对子代成蚜的性状产生正面刺激作用。

总之,母代无论经历高温单作还是高温农药互作,尽管会对子代若虫期(发育和存活)产生负面影响,但这种负面影响却并没有延续到成蚜的生活史性状,甚至会对成蚜寿命产生正面刺激作用。高温类型在与农药互作条件下,对母代和子代寿命、繁殖的影响存在相反的趋势。相较母代高温单作,农药的存在,会对子代成蚜的性状产生正面刺激作用。除极端高温组处理外(38℃/10 min),无论吡虫啉是否存在,随着母代经历高温持续天数的增加,子代成蚜表现均逐渐提升。

母代无论经历高温单作还是高温农药互作,对子代种群内禀增长率均产生的是负面影响。从高温类型对子代麦长管蚜的种群参数的影响可以看出,母代温和高温对子代的总体内禀的负面影响最小。虽然母代38℃/10 min较高高温经历对子代世代时间延长幅度较小,但母代温和高温对子代净繁殖率的正面影响,对子代总体内禀的贡献更大。母代高温与农药互作对子代内禀增长率的影响随着母代高温持续天数的不同而不同。总体上看,随着高温持续天数的增加,高温与农药的互作对子代的 r_m 是有好处的。

第 3 章

第 3 章

高温和吡虫啉胁迫次序对麦长管蚜生活史性状的影响

昆虫在自然界中的栖息环境很少有最适的,通常会暴露在多种胁迫压力之下[97]。其中高温事件的频发就是最常见的自然胁迫,对昆虫的生长、发育、繁殖等基本生命活动均会产生深刻影响[4,43]。而化学农药作为另一种常见的非生物胁迫,同样会改变昆虫的生长轨迹如造成大面积死亡[7],刺激繁殖导致害虫的再猖獗[10]。尽管这两种胁迫均会引起昆虫的表型响应,但是两者性质和作用机制却存在极大差异。

高温对昆虫表型的影响是通过虫体内在一系列的生理生化改变而表现出来的[172],如昆虫暴露在无法逃避的高温环境中,最直接的反应是增加体内水分的蒸发作用来降低机体的温度,避免高温的伤害[173],而体内失水,会导致昆虫体内微环境的改变,离子浓度的增加[174],进而破坏细胞骨架[175],改变细胞膜的完整性和流动性[176]。此外,高温也可以诱导昆虫体内表达热激蛋白,降低高温对昆虫自身蛋白质的破坏,产生保护机制[177]。农药对昆虫的影响,多数是通过毒杀作用,直接作用于昆虫体内的靶标物质,如有机磷类杀虫剂与昆虫体内的乙酰胆碱受体结合抑制乙酰胆碱酯酶的活性[178],拟除虫菊酯类杀虫剂与机体内的受体结合对钠离子通道产生抑制作用[179],新烟碱类杀虫剂作用于虫体内的烟碱乙酰胆碱受体,干扰昆虫的神经系统[180]。

在此背景下,高温与农药两种胁迫出现次序就显得尤为重要,农药与高温的出现次序可能会产生不同的生态表型后果,但这个关键因素却几乎很少受到关注[97]。高温与农药的出现次序可以有三种情况:先药后热、热药同时和先热后药。但生产中为减少农药中毒事故的发生,常常避开中午的高温而在早晨或黄昏用药,因此我们主要开展了先药后热和先热后药两个方向上的交互效应研究。据此,提出了如下问题:

(1)高温和农药先后不同胁迫次序处理对麦长管蚜的存活影响有何差异,是否会影响二者的联合作用?

(2)胁迫次序对个体世代内以及世代间的生活史性状以及整体适合度的影响有何差异?

(3)如果有差异,这一发现有哪些潜在意义?

为了回答上述问题,我们分别考查了高温事件 34℃/180 min 与农药先后处理对麦长管蚜当代存活率、寿命、繁殖以及对子一代的生长、发育繁殖等生活史性状的影响。

◆ 3.1 材料与方法

3.1.1 试验用虫

挑取饲养条件下的新生若蚜约 3 000 头,集中饲养至 9 日后单独分装于饲养管中,用于试验。

3.1.2 供试药剂

95% 吡虫啉原药(中农联合生物科技有限公司提供)。

3.1.3 试验因子设定

3.1.3.1 高温事件

亚致死高温类型设置 34℃/180 min,持续天数为 1 d、3 d、5 d。因此,高温事件设计为 3 种,即 1 种亚致死高温类型 ×3 种持续天数。其中热处理模式仍以 24 h 为周期,每天分别经历 34℃/180 min,一天中剩余时间均设置为 22℃标准恒温,光周期为 L : D=16 : 8。

3.1.3.2 毒苗制备

毒苗制备过程同第 1 章 1.4.2。

3.1.4 试验设计

3.1.4.1 高温和吡虫啉胁迫次序对麦长管蚜即时存活率联合作用测定

开展了高温农药吡虫啉先后不同胁迫次序互作对麦长管蚜存活的联合作用测定,其中高温处理周期为 1 d、3 d 和 5 d;农药吡虫啉处理时间为 1 d,但有 3 个施药时间,分别为第 1 天、第 3 天以及第 5 天(图 3-1)。编号 P_1st 代表农药单作(P 为 pesticide,1st 代表在第一天施药,H_1 代表高温单作(H 为 heat,1 为高温处理持续 1 d),HP_1 代表先高温后农药处理(H 为 heat,P 为 pesticide,1 同样为高温处理持续天数 1 d),PH_1 代表先农药后高温处理(P 为 pesticide,H 为 heat,1 同样为高温处理持续天数 1 d)。整个试验由 3 个高温单作(1 种高温类型 ×3 种持续天数)+6 个高温和农药互作(3HP+3PH)+3 个 22℃农药单作(P)+1 个 22℃空白对照(CK)=13 个处理。

先热后药处理(HP):9 日龄麦长管蚜,每 10 头置于 1 支 1.5 mL 离心管中,管中放有新鲜麦苗,离心管插入带孔泡沫板固定后,置于培养箱中先进行 3 种高温事件处理(HP_1、HP_3、HP_5),高温处理结束后,用毒苗替换管中新鲜麦苗,转移到恒温

22 ℃下继续进行农药处理,1 d 后,调查死亡数。试验期间麦苗每 3 天更换 1 次。

先药后热处理(PH):9 日龄麦长管蚜,每 10 头置于 1 支 1.5 mL 离心管中,管中放有毒苗,置于 22 ℃先进行农药处理,1 d 后,用新鲜麦苗替换管中的毒苗,置于培养箱中进行 3 种高温事件处理(PH_1、PH_3、PH_5),处理结束后 1 d 调查死亡数。

高温单作(H):9 日龄麦长管蚜,每 10 头置于 1 支 1.5 mL 离心管中,管中放有新鲜麦苗,离心管插入带孔泡沫板固定后,置于培养箱中进行 3 种高温事件处理(H_1、H_3、H_5),处理结束后 1 d 调查死亡数。试验期间麦苗每 3 天更换 1 次。

农药对照(P):由于先热后药,药剂处理的时间不同,因此,农药对照也设置了相应的处理时间,即在试验开始的第 1 天(P_1st),第 3 天(P_3rd),第 5 天(P_5th)分别进行农药施用,每 10 头置于 1 支 1.5 mL 离心管中,管中放有毒苗,置于恒温 22 ℃下,1 d 后调查死亡数。此外,由于先药后热各处理结束后时间差距较大,因此我们同时也调查了第 1 天施药(P_1st),1 d 后转移到新鲜麦苗后,分别于第 3 天、第 5 天、第 7 天调查死亡数,以作为先药后热各处理单独药后的死亡率。

空白对照(CK):由于该试验各处理结束后时间相差太大(最短为 3 h,最长为 6 d),所以分别设置了各自的空白对照,即将 9 日龄麦长管蚜,每 10 头置 1 支 1.5 mL 离心管中,管中放有新鲜麦苗,置于恒温 22 ℃下,分别于第 2 天、第 3 天、第 4 天、第 5 天、第 6 天、第 7 天均调查了死亡数。试验期间麦苗每 3 天更换 1 次。

以上所有处理均设置 4 次重复,每重复 10 头蚜虫。

3.1.4.2 高温和吡虫啉胁迫次序对麦长管蚜生态性状的影响

为了研究高温农药吡虫啉先后不同胁迫次序互作对麦长管蚜当代以及跨代生活史参数的影响,设计了高温农药先后顺序互作的试验,包括 1 个高温类型(34 ℃/180 min),3 个高温持续天数(1 d、3 d、5 d),2 个农药胁迫次序处理(先后),同时增加了 1 个空白对照了用于对比高温和农药对生活史性状的实际影响,共计 10 个处理(图 3-1)。其中 60 头蚜虫/处理用于母代生活史性状调查;另 60 头蚜虫 / 处理仅用于子一代采集。

高温单作(H):将 9 日龄麦长管蚜单头接入插有新鲜无药麦苗的饲养管后,置于培养箱中进行 34℃/180 min 的高温处理,持续天数分别为 1 d、3 d、5 d,处理结束后将蚜虫转移到恒温 22℃下继续饲养,试验期间麦苗每 3 天更换 1 次。

先高温后农药(HP):将 9 日龄麦长管蚜单头接入插有新鲜无药麦苗的饲养管后,置于培养箱中进行 34℃/180 min 的高温处理,持续天数分别为 1 d、3 d、5 d,处理结束后将蚜虫转移到插有毒苗的饲养管中,置于 22℃进行农药处理,处理 1 d 后,转移到新鲜麦苗的离心管中,继续放置于恒温 22℃环境下,直至供试蚜虫全部死亡。试验期间麦苗每 3 天更换 1 次。

先农药后高温(PH):将 9 日龄麦长管蚜单头接入插有毒苗的饲养管后,置于 22℃进行农药处理,处理 1 d 后,转移到新鲜麦苗的饲养管中,置于培养箱中进行 34℃/180 min 的高温处理,持续天数分别为 1 d、3 d、5 d,处理结束后,转移到恒温 22℃下继续饲养,直至供试蚜虫全部死亡。试验期间麦苗每 3 天更换 1 次。注:因为空白对照第二天的死亡率仍为 0,所以计算先农药后高温(PH)各处理的理论死亡率仍然按 H_{-1}、H_{-3} 和 H_{-5} 的死亡率作为单独热导致的死亡率。

空白对照(CK):将 9 日龄麦长管蚜单头接入插有新鲜无药麦苗的饲养管后,置于恒温 22℃下一直饲养,直至供试蚜虫全部死亡。

母代性状调查:试验期间,每天都调查记录 1 次存活状态、产仔情况,并将计数后若蚜去除,直至供试蚜虫全部死亡,母代性状调查结束。存活寿命从第 9 天开始算起,直到死亡的时间;繁殖同样从第 9 天开始记录,整个存活时期每头蚜虫所产后代的总数。繁殖率是指每头蚜虫在成蚜阶段内平均每天所产的后代数即繁殖/寿命。

跨代性状调查:所有处理结束后第二天调查时,随机采集 30 头子代 / 处理(共 10 个处理)若蚜置于常温 22℃养虫室内,于每日上午 8 点计数死亡个体、蜕皮及产仔情况,并将蜕皮、死亡蚜虫及新生若蚜去除,直至供试蚜虫全部死亡。去除逃逸蚜虫,所有处理蚜虫均测定了以下指标:若虫死亡率、发育历期、成蚜产仔前期、成蚜寿命以及繁殖。若虫死亡率是指活到成蚜的若虫占全部测试若虫的比例。成蚜寿命是指从变成成蚜到蚜虫死亡的时间。成蚜繁殖是指每头成蚜所产后代的总数。繁殖率是指每头蚜虫在成蚜阶段内平均每天所产的后代数也即成蚜繁殖/成蚜寿命。

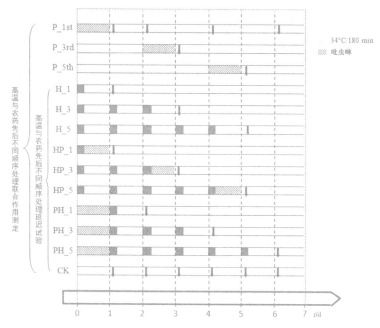

图3-1 高温农药吡虫啉胁迫次序互作流程图。"↓"代表每处理的死亡率调查点以及子代采样点。

注：空白对照子代采样点为第一个"↓"代表的时刻。第0天被定义为实验开始的时间。

Fig.3-1 Experimental design of heat event interaction with imidacloprid in different sequences. ↓ indicates mortality assessment and offsprings sampling points for each treatment.

Note: The first ↓ indicate offspring sampling point for CK. "0" indicated when the experiment started.

3.1.5 统计分析

协同毒力指数 = (校正死亡率 − 理论死亡率)/理论死亡率 × 100

其中,协同毒力指数≥20为协同作用,协同毒力指数≤-20为拮抗作用,处于二者之间为相加作用。

其中:

理论死亡率 = 单独热校正后的死亡率 + 单独药校正后的死亡率 - 两者的乘积

校正死亡率 = (未处理对照组生存率 - 处理组生存率)/未处理对照组生存率 × 100

母代繁殖、寿命、繁殖率以及子代发育历期和产仔前期性状,在分析之前,需经平方根转换后提高其正态性。转换后,数据统计分析均使用转换后的,为了作图的直观仍用原始数据。母代繁殖、寿命、繁殖率和子代发育历期、产仔前期、成蚜寿命、成蚜繁殖性状均采用一般线性模型(GLM)进行方差分析,利用Duncan法进行处理间多重比较,利用独立样本t检验来比较高温农药两胁迫次序处理间的差异显著性(SPSS 19.0)。此外,所有处理和空白对照间的差异显著性比较采用Dunnett-t检验法(SPSS 19.0)。子代若蚜存活率采用列联表法进行存活率显著性分析,非独立2×2以及2×3表法实现处理间的多重比较(SPSS 19.0)。

种群参数计算与分析同第1章1.6种群参数的统计分析。

◆ 3.2 结果与分析

3.2.1 高温和吡虫啉胁迫次序对即时死亡率联合作用评价

由表3-1可知,高温农药先后不同胁迫次序对麦长管蚜所产生的联合毒力结果不尽相同。在热持续1 d下,先高温后农药处理时,协同毒力指数为-46.67,表现为拮抗作用,但先农药后高温处理,协同毒力指数为57.01,表现为协同作用;在热持续3 d下,先高温后农药和先农药后高温处理的协同毒力指数分别为14.62和7.18,均表现为相加作用;同样的,在热持续5 d下,先高温后农药和先农药后高温处理的协同毒力指数分别为-3.80和-2.16,均表现为相加作用。

表 3-1 高温和吡虫啉胁迫次序对麦长管蚜即时死亡率联合作用评价

Table 3-1 Assessment of the effects of heat event interaction with imidacloprid in different sequences on immediate mortality of *Sitobion avenae*

处理 Treatment	温度 Temperature /℃	热持续时间 Duration time /d	吡虫啉处理与时间点 Application concentration and time		调查时间 Time	死亡率 Mortality			协同毒力指数 Cooperative virulence index (c.f.)	结果评定 Results assessment
						实际值 Real	校正值 Corrected	理论值 Theoretical		
CK	22	0	0		2nd	0	0			
					3rd	0.05	0.05			
					4th	0.07	0.07			
					5th	0.13	0.13			
					6th	0.22	0.22			
					7th	0.27	0.27			
P	22	0	20	1st	2nd	0.25	0.25			
					3rd	0.42	0.39			
					5th	0.73	0.69			
					7th	0.92	0.89			
			20	3rd	4th	0.32	0.27			
			20	5th	6th	0.48	0.34			

续表

处理 Treatment	温度 Temperature /°C	热持续时间 Duration time /d	吡虫啉处理与时间点 Application concentration and time		调查时间 Time	死亡率 Mortality			协同毒力指数 Cooperative virulence index (c.f.)	结果评定 Results assessment
						实际值 Real	校正值 Corrected	理论值 Theoretical		
H	34	1	0		2nd	0.17	0.17			
	34	3	0		4th	0.47	0.43			
	34	5	0		6th	0.80	0.74			
HP	34	1	20	1st	2nd	0.20	0.20	0.38	−46.67	拮抗作用
	34	3	20	3rd	4th	0.67	0.64	0.58	14.62	相加作用
	34	5	20	5th	6th	0.80	0.74	0.83	−3.80	相加作用
PH	34	1	20	1st	3rd	0.77	0.75	0.49	57.01	协同作用
	34	3	20	1st	5th	0.88	0.87	0.82	7.18	相加作用
	34	5	20	1st	7th	0.95	0.93	0.97	−2.16	相加作用

3.2.2 高温和吡虫啉胁迫次序对当代生活史性状的影响

3.2.2.1 寿命

不同持续天数高温事件和吡虫啉不同处理对麦长管蚜当代寿命的两因子方差分析结果表明：吡虫啉不同处理对当代麦长管蚜寿命影响差异显著，但不同高温事件及其二者交互作用对当代寿命的影响不显著（表 3-2）。

表 3-2 高温和吡虫啉不同胁迫处理对麦长管蚜当代寿命的方差分析

Table 3-2 Results of variance analysis for effects of heat event interaction with imidacloprid in different treatments on maternal longevity of *S.avenae*

来源 Source	DF	MS	F	P
高温事件 Heat wave	2	0.92	1.58	0.206
吡虫啉处理 Imidacloprid treatment	2	20.11	34.60	<0.001
高温事件 × 吡虫啉处理 Heat wave × Imidacloprid treatment	4	1.20	2.07	0.084
误差 Error	531	—	—	—

与空白对照寿命（13.58±0.99）d 相比，所有处理均显著缩短了当代麦长管蚜的寿命（图 3-2），最少缩短了（6.58±1.14）d。且高温农药两胁迫互作对当代寿命产生的负面影响均显著超过了仅高温处理所带来的负面效应，无论是在热持续 1 d（$F_{2,177}$ = 12.916, P <0.001）、3 d（$F_{2,177}$ = 16.924, P <0.001）还是 5 d（$F_{2,177}$ = 5.248, P =0.006），但两胁迫次序却并未对该性状产生显著影响（图 3-2）。此外，当两胁迫同时存在时，热持续时间未对寿命产生显著影响，无论是先高温后农药（$F_{2,177}$ = 0.179, P =0.836）还是先农药后高温（$F_{2,177}$ = 0.122, P =0.885），但当仅存在高温胁迫时，热持续天数的增加却对该性状产生了影响，且随着持续天数的增加产生的负面影响增大（$F_{2,177}$ = 3.250, P =0.041）。

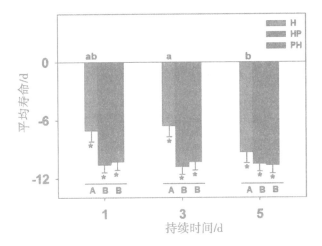

图 3-2 高温和吡虫啉不同胁迫处理对当代麦长管蚜寿命的影响（平均值 ± 标准误）。每个柱子表示处理组减去空白对照组对应的平均值与标准差。"*"表示处理组和空白对照组间差异达显著水平（$P=0.05$）。大写字母代表不同色柱处理方式间的差异显著性（$P=0.05$）。小写字母代表同色柱不同持续天数高温事件间的差异显著性（$P=0.05$）。H、HP、PH 代表不同胁迫处理，H 为仅高温单作，HP 为先高温后农药，PH 为先农药后高温，下同之。

Fig. 3-2 Effects of heat event interaction with imidacloprid in different treatments on maternal longevity (mean ± SE) of S.avenae. Each bar represents the mean value and standard error corresponding to the different treatment minus the CK treatment. The "*" represent significant difference ($P=0.05$) between treatment and CK group. Different capital letters indicate significant level ($P=0.05$) of values between different treatments. Different lowercase letters indicate significant level ($P=0.05$) of values between heat wave with different durations. H, HP and PH represent different treatments. H is only heat wave. HP is treated with heat wave first, then applied imidacloprid. PH is applied imidacloprid first, then treated with heat wave.

3.2.2.2 繁殖

不同持续天数高温事件和吡虫啉不同处理对麦长管蚜当代繁殖的两因子方差分析结果表明：不同持续天数高温事件、吡虫啉不同处理以及二者交互均对当代麦长管蚜的繁殖影响差异显著(表3-3)。

表3-3 高温和吡虫啉不同胁迫处理对麦长管蚜当代繁殖的方差分析

Table 3-3 Results of variance analysis for effects of heat event interaction with imidacloprid in different treatments on maternal fecundity of *S.avenae*

来源 Source	DF	MS	F	P
高温事件 Heat wave	2	4.30	3.17	0.043
吡虫啉处理 Imidacloprid treatment	2	60.99	45.00	<0.001
高温事件 × 吡虫啉处理 Heat wave × Imidacloprid treatment	4	4.39	3.24	0.012
误差 Error	531	1.36	—	—

与寿命影响类似,所有处理均显著压低了当代麦长管蚜的繁殖,相较空白对照的成蚜繁殖力(15.28±1.59)头若蚜/成蚜(图3-3),最少压低了(7.97±1.55)头。且高温农药两胁迫互作对当代繁殖产生的负面影响均显著超过了仅高温处理所带来的负面效应,但无论是在热持续1 d ($F_{2,177}$ = 26.122, P <0.001)、3 d ($F_{2,177}$ = 13.683, P <0.001)还是5 d ($F_{2,177}$ = 7.049, P =0.001),两胁迫次序均未对该性状产生显著影响(图3-3)。此外,当两胁迫同时存在时,热持续时间未对寿命产生显著影响,无论是先高温后农药($F_{2,177}$ = 0.413, P =0.662)还是先农药后高温($F_{2,177}$ = 1.887, P =0.155),但当仅存在高温胁迫时,热持续天数的增加却对该性状产生影响($F_{2,177}$ = 5.296, P =0.006)。

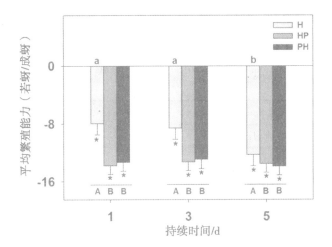

图 3-3 高温农药吡虫啉不同处理下当代麦长管蚜繁殖的变化（平均值 ± 标准误）。每个柱子表示处理组减去空白对照组对应的平均值与标准差。"*" 表示处理组和空白对照组间差异达显著水平（$P=0.05$）。大写字母代表不同色柱处理方式间的差异显著性（$P=0.05$）。小写字母代表同色柱不同持续天数高温事件间的差异显著性（$P=0.05$）。

Fig. 3-3 Effects of heat event interaction with imidacloprid in different treatments on maternal fecundity (mean ± SE) of *S. avenae*. Each bar represents the mean value and standard error corresponding to the different treatment minus the CK treatment. The "*" represent significant difference ($P=0.05$) between treatment and CK group. Different capital letters indicate significant level ($P=0.05$) of values between different treatments. Different lowercase letters indicate significant level ($P=0.05$) of values between heat wave with different durations.

3.2.2.2 繁殖率

不同持续天数高温事件和吡虫啉不同处理对麦长管蚜当代繁殖的两因子方差分析结果表明：不同持续天数高温事件对当代麦长管蚜的繁殖率影响差异不显著，但吡虫啉不同处理以及二者交互均会对该性状产生显著影响（表 3-4）。

表3-4 高温农药吡虫啉不同胁迫处理对麦长管蚜当代繁殖率的方差分析

Table 3-4 Results of variance analysis for effects of heat event interaction with imidacloprid in different treatments on maternal fecundity rate of S.avenae

来源 Source	DF	MS	F	P
高温事件 Heat wave	2	0.48	2.43	0.089
吡虫啉处理 Imidacloprid treatment	2	6.48	32.69	<0.001
高温事件 × 吡虫啉处理 Heat wave × Imidacloprid treatment	4	0.55	2.79	0.026
误差 Error	531	—	—	—

与空白对照每天繁殖率（0.97±0.06）头若蚜/成蚜相比，所有处理均压低了当代麦长管蚜的繁殖率（图3-4），每天最少压低了（0.07±0.09）头若蚜/成蚜。且高温农药两胁迫互作对当代寿命产生的负面影响均显著超过了仅高温处理所带来的负面效应，无论是在热持续1 d（$F_{2,177}$ = 26.366, P <0.001）、3 d（$F_{2,177}$ = 5.973, P =0.003）还是5 d（$F_{2,177}$ = 6.853, P =0.001），且两胁迫次序仅在热持续1 d下对该性状产生显著影响（图3-4）。此外，当两胁迫同时存在时，热持续时间仅在先药后热胁迫次序下对繁殖率产生显著影响（$F_{2,177}$ = 6.365, P =0.002）。

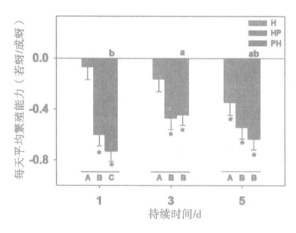

图3-4 高温和吡虫啉不同胁迫处理对当代麦长管蚜繁殖率的影响（平均值 ± 标准误）。每个柱子表示处理组减去空白对照组对应的平均值与标准

差。"*"表示处理组和空白对照组间差异达显著水平（P=0.05）。大写字母代表不同色柱处理方式间的差异显著性（P=0.05）。小写字母代表同色柱不同持续天数高温事件间的差异显著性（P=0.05）。

Fig. 3-4 Effects of heat event interaction with imidacloprid in different treatments on maternal fecundity rate（mean ± SE）of *S.avenae*. Each bar represents the mean value and standard error corresponding to the different treatment minus the CK treatment. The "*" represent significant difference （P=0.05）between treatment and CK group. Different capital letters indicate significant level（P=0.05）of values between different treatments. Different lowercase letters indicate significant level（P=0.05）of values between heat wave with different durations.

3.2.2.3 种群参数

与空白对照麦长管蚜的种群参数相比，经高温农药不同处理后虽然平均世代时间（G）缩短[图 3-5(C)]，但由于种群净繁殖率（R_0）显著下降[图 3-5(B)]，整体的内禀增长率（r_m）仍然显著下降[图 3-5(A)]。高温农药两胁迫互作对当代种群参数产生的影响均显著超过了仅高温处理，且两胁迫次序也对种群参数产生了显著影响（P<0.05，图 3-5），如 HP 相较 PH 显著的缩短种群平均世代时间；在热持续 1 d 和 3 d 的情况下，HP 相较 PH 显著地降低了种群的内禀增长率和净繁殖率，但在热持续 5 d，却产生了与之相反的效应（图 3-5）。此外，当仅存在高温单作时，随着高温持续时间的增加，种群内禀增长率（r_m）、净繁殖率（R_0）显著下降，平均世代时间（G）先增加后缩短（图 3-5）。在先高温后农药（HP）处理下，随着高温持续时间的增加，种群内禀增长率（r_m）、净繁殖率（R_0）显著升高[图 3-5A 和 (C)]，平均世代时间（G）显著缩短[图 3-5(B)]。而在先农药后高温（PH）处理下，随着高温持续时间的增加，种群内禀增长率（r_m）、净繁殖率（R_0）先增加后下降[图 3-5(A)、(C)]，平均世代时间（G）显著缩短[图 3-5(B)]。

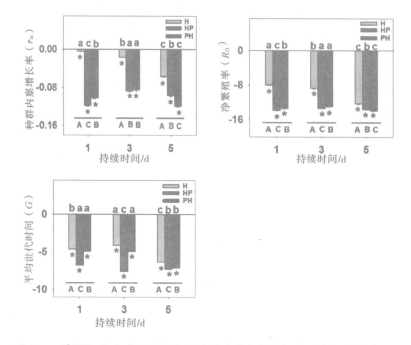

图 3-5 高温和吡虫啉不同胁迫处理对当代麦长管蚜种群参数的影响（平均值 ± 标准误）。每个柱子表示处理组减去空白对照组对应的平均值与标准差。"*"表示处理组和空白对照组间差异达显著水平（$P=0.05$）。大写字母代表不同色柱处理方式间的差异显著性（$P=0.05$）。小写字母代表同色柱不同持续天数高温事件间的差异显著性（$P=0.05$）。

Fig. 3-5 Effects of heat event interaction with imidacloprid in different treatments on maternal population parameters (mean ± SE) of *S. avenae*. Each bar represents the mean value and standard error corresponding to the different treatment minus the CK treatment. The "*" represent significant difference ($P=0.05$) between treatment and CK group. Different capital letters indicate significant level ($P=0.05$) of values between different treatments. Different lowercase letters indicate significant level ($P=0.05$) of values between heat wave with different durations.

3.2.3 高温和吡虫啉胁迫次序对跨代生活史性状的影响

3.2.3.1 子代发育历期

表 3-5 母代经历高温农药不同胁迫处理对子代若蚜发育历期的方差分析

Table 3-5 Results of variance analysis for effects of maternal exposing heat event interaction with imidacloprid in different treatments on offspring nymphal time of *S.avenae*

母代处理 Maternal treatment	DF	MS	F	P
高温事件 Heat wave	2	0	0.11	0.893
吡虫啉处理 Treatment	2	0.1	3.73	0.026
高温事件 × 吡虫啉处理 Heat wave × Imidacloprid treatment	4	0.02	0.55	0.699
误差 Error	161	0.03	—	—

母代经历不同热持续天数高温事件和农药吡虫啉不同处理对子代若蚜发育历期的两因子方差分析结果表明：母代经历农药不同处理对子代若蚜发育历期影响差异显著，但母代经历不同热持续天数高温事件及二者交互对该性状影响不显著（表 3-5）。

与空白对照所产子代的发育历期（8.72±0.22）d 相比，母代经所有处理后均不同程度的延长了子代发育所需时间，但仅在母代先经 5 d 高温后经农药处理子代若蚜发育所需时间最长为（9.63±0.301）d 且达到了显著水平（图 3-6 P=0.032）；相较母代经高温对照处理所产子代的发育历期，母代经先高温后农药处理延长了子代发育所需时间，而经先农药后高温处理却缩短了子代的发育历期（图 3-6），同时母代经高温农药先后次序处理在热持续 1 d（t = 2.033, df = 36, P = 0.050）和 5 d（t = 2.368, df = 29, P = 0.025）也会对子代的发育历期产生显著差异（图 3-6）。

3.2.3.2 子代若虫存活率

与空白对照所产子代存活率 93.3% 相比，母代经所有处理后均不同程度的压低了子代的若虫存活率，且空白与 PH_1（χ^2 = 6.667, df = 1, P = 0.010）、HP_3（χ^2 = 4.320, df = 1, P = 0.038）、PH_3（χ^2 = 10.756, df = 1, P = 0.001）、HC_5（χ^2 = 4.584, df =

1, $P = 0.032$）、HP_5（$\chi^2 = 12.273$, $df = 1$, $P < 0.001$）、PH_5（$\chi^2 = 13.871$, $df = 1$, $P < 0.001$）处理相比均达到了显著水平（图 3-7）。无论热持续几天，相较母代经高温单作后子代的存活率，母代经高温农药互作后均降低了子代的若虫存活率，与此同时先农药后高温处理所产生的负面影响最大（图 3-7），分别是热持续 1 d 的 66.7%，热持续 3 d 的 56.7% 以及热持续 5 d 的 50%。此外，无论母代是经高温单作处理还是高温农药互作处理，随着热持续天数的增加，对子代若虫存活的影响越大，甚至在先高温后农药处理下，不同热持续天数所产生的影响达到了临界显著水平（$\chi^2 = 5.392$, $df = 2$, $P = 0.067$）。

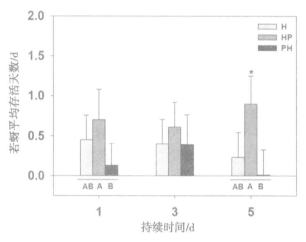

图 3-6 母代经历高温和吡虫啉不同胁迫处理后对子代若蚜历期的影响（平均值 ± 标准误）。每个柱子表示处理组减去空白对照组对应的平均值与标准误。"*" 表示处理组和空白对照组间差异达显著水平（$P=0.05$）。大写字母代表不同色柱处理方式间的差异显著性（$P=0.05$）。小写字母代表同色柱不同持续天数高温事件间的差异显著性（$P=0.05$）。无标记则为差异不显著。

Fig. 3-6 Effects of maternal exposing heat event interaction with imidacloprid in different treatments on offspring nymphal time（mean ± SE）of S.avenae. Each bar represents the mean value and standard error corresponding to the different treatment minus the CK treatment. The "*" represent significant difference（$P=0.05$）between treatment and CK group. Different capital letters indicate significant level（$P=0.05$）of values between different treatments. Different lowercase letters indicate significant level

(P=0.05) of values between heat wave with different durations. No mark means the difference is not significant.

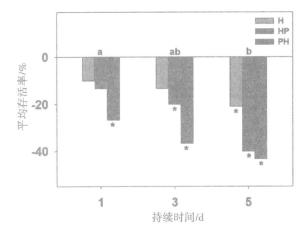

图 3-7 母代经历高温和吡虫啉不同胁迫处理后对子代若虫存活率的影响（平均值 ± 标准误）。每个柱子表示处理组减去空白对照组对应的平均值。"*"表示处理组和空白对照组间差异达显著水平（P=0.05）。小写字母代表同色柱不同持续天数高温事件间的差异显著性（P=0.05）。大写字母代表不同色柱处理方式间的差异显著性（P=0.05）。小写字母代表同色柱不同持续天数高温事件间的差异显著性（P=0.05）。无标记则为差异不显著。

Fig. 3-7 Effects of maternal exposing heat event interaction with imidacloprid in different treatments on on offspring survival rate (mean ± SE) of *S.avenae*. Each bar represents the mean value and standard error corresponding to the different treatment minus the CK treatment. The "*" represent significant difference (P=0.05)between treatment and CK group. Different capital letters indicate significant level (P=0.05)of values between different treatments. Different lowercase letters indicate significant level (P=0.05) of values between heat wave with different durations. No mark means the difference is not significant.

3.2.3.3 子代产仔前期

母代经历不同热持续天数高温事件和吡虫啉不同处理对子代成蚜产仔前期的两因子方差分析结果表明：母代经历吡虫啉不同处理以及不同热持续天数高温事件对子代成蚜产仔前期影响均不

差异显著,但二者交互却对该性状产生临界显著影响(表3-6)。

表3-6 母代经历高温和吡虫啉不同胁迫处理对子代麦长管蚜产仔前期的方差分析

Table 3-6 Results of variance analysis for effects of maternal exposing heat event interaction with imidacloprid in different treatments on offspring pre-productive period of S.avenae

母代处理 Maternal treatment	DF	MS	F	P
高温事件 Heat wave	2	0.06	0.19	0.826
吡虫啉处理 Treatment	2	0.59	1.98	0.141
高温事件 × 吡虫啉处理 Heat wave × Imidacloprid treatment	4	0.71	2.39	0.053
误差 Error	161	0.3	—	—

与空白对照所产子代的产子前期(0.47 ± 0.11)d相比,母代经单热1 d、3 d以及先农药后热1 d处理后子代的产子前期缩短了(0.02 ± 0.15)d到(0.13 ± 0.16)d,除此之外,其余处理均不同程度的延长了后代的产子前期(0.03 ± 0.19)d到(0.39 ± 0.20)d,但均未达到显著水平(图3-8);母代经高温单作、高温与农药不同胁迫次序互作处理,对子代产子前期的影响在热持续3 d($F_{2,54}= 1.062, P =0.353$)、5 d($F_{2,49}= 1.369, P =0.264$)均没有产生显著影响,仅在热持续1 d产生了显著影响($F_{2,58}= 4.962, P =0.010$),且两胁迫次序同样对子代产子前期也产生了显著影响($t = 3.098, df = 25.011, P = 0.005$)(图3-6)。此外,母代经历高温农药互作,无论是先高温后农药($F_{2,49}= 1.392, P =0.258$)还是先农药后高温($F_{2,47}= 0.603, P =0.551$),热持续时间均未对子代产子前期产生显著影响,但当母代仅经历高温单作处理($F_{2,65}= 3.496, P =0.036$),热持续天数就会对子代产子前期产生显著影响(图3-8)。

3.2.3.4 子代成蚜寿命

母代经历不同热持续天数高温事件和吡虫啉不同处理对子代成蚜寿命的两因子方差分析结果表明:母代经历吡虫啉不同处理对子代成蚜寿命影响差异显著,但母代经历不同热持续天数高温事件及二者交互对该性状影响不显著(表3-7)。

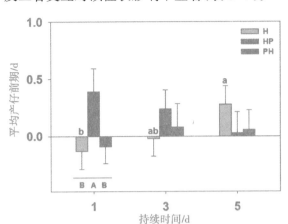

图3-8 母代经历高温和吡虫啉不同胁迫处理后对子代成蚜产仔前期的影响(平均值 ± 标准误)。每个柱子表示不同处理组减去空白对照组对应的平均值与标准误。"*"表示处理组和空白对照组间差异达显著水平($P=0.05$)。大写字母代表不同色柱处理方式间的差异显著性($P=0.05$)。小写字母代表同色柱不同持续天数高温事件间的差异显著性($P=0.05$)。无标记则为差异不显著。

Fig. 3-8 Effects of maternal exposing heat event interaction with imidacloprid in different treatments on offspring pre-productive period (mean ± SE) of *S.avenae*. Each bar represents the mean value and standard error corresponding to the different treatment minus the CK treatment. The "*" represent significant difference ($P=0.05$)between treatment and CK group. Different capital letters indicate significant level ($P=0.05$)of values between different treatments. Different lowercase letters indicate significant level ($P=0.05$)of values between heat wave with different durations. No mark means the difference is not significant.

表3-7 母代经历高温和吡虫啉不同胁迫处理对子代成蚜寿命的方差分析

Table 3-7 Results of variance analysis for effects of maternal exposing heat event interaction with imidacloprid in different treatments on offspring longevity of *S.avenae*

母代处理 Maternal treatment	DF	MS	F	P
高温事件 Heat wave	2	7.26	0.17	0.845

续表

母代处理 Maternal treatment	DF	MS	F	P
吡虫啉处理 Treatment	2	736.02	17.12	<0.001
高温事件 × 吡虫啉处理 Heat wave × Imidacloprid treatment	4	45.66	1.06	0.377
误差 Error	161	43	—	—

与空白对照所产子代的寿命（13.68 ± 1.40）d 相比，除母代经单热 1 d、3 d（HC_1、HC_3）对子代寿命有负面影响外，其余处理均对子代的寿命产生了刺激作用，且在母代经 PH_1（$t = -2.714, df = 49, P = 0.009$）和 PH_3（$t = -3.195, df = 32, P = 0.003$）处理下达到了显著水平（图 3-9）。无论热持续几天，相较母代经高温单作后子代成蚜的寿命，母代经高温农药互作后均刺激了子代成蚜的寿命，同时在热持续 1 d（$F_{2,58}= 6.108, P =0.004$）和 3 d（$F_{2,54}= 10.940, P <0.001$）达到了显著水平，且两胁迫次序处理同样会对子代成蚜寿命产生影响，相较先高温后农药（HP）处理后的子代成蚜寿命，先农药后高温（PH）对子代寿命的刺激作用更强烈（图 3-9）。

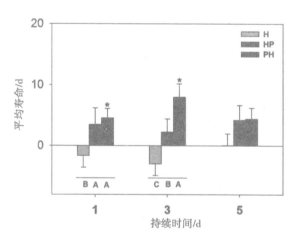

图 3-9 母代经历高温和吡虫啉不同胁迫处理后对子代成蚜寿命的影响。每个柱子表示处理组减去空白对照组对应的平均值与标准误。"*"表示处理组和空白对照组间差异达显著水平（$P=0.05$）。大写字母代表不同色柱处理方式间的差异显著性（$P=0.05$）。小写字母代表同色柱不同持续天数高温事件间的差异显著性（$P=0.05$）。无标记则为差异不显著。

Fig. 3-9 Effects of maternal exposing heat event interaction with imidacloprid in different treatments on offspring nymphal time (mean ± SE) of S.avenae. Each bar represents the mean value and standard error corresponding to the different treatment minus the CK treatment. The "*" represent significant difference ($P=0.05$) between treatment and CK group. Different capital letters indicate significant level ($P=0.05$) of values between different treatments. Different lowercase letters indicate significant level ($P=0.05$) of values between heat wave with different durations. No mark means the difference is not significant.

3.2.3.5 子代成蚜繁殖

母代经历不同热持续天数高温事件和吡虫啉不同处理对子代成蚜繁殖的两因子方差分析结果表明：母代经历吡虫啉不同处理对子代成蚜繁殖影响差异显著，但母代经历不同热持续天数高温事件及二者交互对该性状影响不显著（表3-8）。

表 3-8 母代经历高温和吡虫啉不同胁迫处理对子代成蚜繁殖的方差分析

Table 3-8 Results of variance analysis for effects of maternal exposing heat event interaction with imidacloprid in different treatments on offspring fecundity of S.avenae

母代处理 Maternal treatment	DF	MS	F	P
高温事件 Heat wave	2	350.15	1.22	0.299
吡虫啉处理 Treatment	2	3486.02	12.1	<0.001
高温事件 × 吡虫啉处理 Heat wave × Imidacloprid treatment	4	264.15	0.92	0.456
误差 Error	158	288.01	—	—

与空白对照所产子代的繁殖力（34.83±3.27）头若蚜/成蚜相比，除母代经单热 1 d、3 d（HC_1、HC_3）对子代繁殖有负面影响外，其余处理均对子代的繁殖产生了刺激作用，且在母代经 PH_3（$t = -3.215, df = 31, P = 0.003$）和 PH_5（$t = -2.534, df = 37, P = 0.016$）处理下达到了显著水平（图3-10）。无论热持续几天，相较母代经高温单作后子代成蚜的繁殖力，母代经

高温农药互作后均刺激了子代成蚜的繁殖,同时在热持续 1 d ($F_{2,57}$= 4.294, P =0.018)和 3 d ($F_{2,52}$= 8.174, P =0.001)达到了显著水平,且两胁迫次序处理同样会对子代成蚜繁殖产生影响,相较先高温后农药(HP)处理后的子代成蚜繁殖,先农药后高温(PH)对子代繁殖的刺激作用更强烈(图 3-10)。

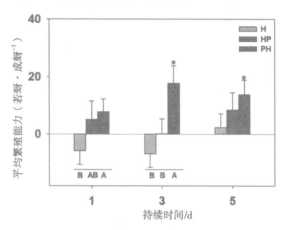

图 3-10 母代经历高温和吡虫啉不同处理后对子代成蚜繁殖的影响(平均值 ± 标准误)。每个柱子表示处理组减去空白对照组对应的平均值与标准误。"*"表示处理组和空白对照组间差异达显著水平(P=0.05)。大写字母代表不同色柱处理方式间的差异显著性(P=0.05)。小写字母代表同色柱不同持续天数高温事件间的差异显著性(P=0.05)。无标记则为差异不显著。

Fig. 3-10 Effects of maternal exposing heat event interaction with imidacloprid in different treatments on offspring nymphal time (mean ± SE) of S.avenae. Each bar represents the mean value and standard error corresponding to the different treatment minus the CK treatment. The "*" represent significant difference (P=0.05) between treatment and CK group. Different capital letters indicate significant level (P=0.05) of values between different treatments. Different lowercase letters indicate significant level (P=0.05) of values between heat wave with different durations. No mark means the difference is not significant.

3.2.3.6 子代成蚜繁殖率

母代经历不同热持续天数和吡虫啉不同处理对子代成蚜繁

殖率的两因子方差分析结果表明：母代经历吡虫啉不同处理对子代成蚜繁殖率影响临界显著,但母代经历高温事件持续天数及二者交互作用不显著(表3–9)。

表3–9 母代经历高温和吡虫啉不同胁迫处理对子代成蚜繁殖率的方差分析

Table 3–9 Results of variance analysis for effects of maternal exposing heat event interaction with imidacloprid in different treatments on offspring fecundity rate of S.avenae

母代处理 Maternal treatment	DF	MS	F	P
高温事件 Heat wave	2	0.36	0.92	0.402
吡虫啉处理 Treatment	2	1	2.52	0.083
高温事件 × 吡虫啉处理 Heat wave × Imidacloprid treatment	4	0.4	1.01	0.402
误差 Error	157	0.4	—	—

与空白对照每天所产子代的成蚜繁殖率(2.51 ± 0.15)头若蚜/成蚜相比,除母代经 PH_3、HC_5 以及 PH_5 处理后每天所产子代成蚜繁殖率增长了(0.00 ± 0.26)头/若蚜/成蚜到(0.19 ± 0.20)头若蚜/成蚜,其余处理均产生了负面影响,但均未达到显著水平(图3–11)。此外,母代经高温单作和高温农药不同胁迫次序互作仅在热持续5 d对子代繁殖率产生了临界显著影响(图3–11; $F_{2,48}= 2.968, P =0.061$)。

母代经历高温农药不同胁迫次序处理后对子代种群参数产生了复杂的影响,既有刺激作用也有抑制作用(图3–11)。对于子代种群内禀增长率 r_m 而言,与空白对照所产子代的 r_m 相比,只有母代经 PH_1 和 PH_5 处理后,不同程度地刺激了子代的 r_m 外,母代经历其他处理均不同程度地降低了子代的 r_m [图3–11(A)]。母代经高温单作和高温农药不同胁迫次序互作,无论在热持续几天下均对子代的 r_m 产生了显著影响($P<0.05$),且在热持续1 d 和 5 d 下,母代经先农药后高温(PH)处理后子代的 r_m 均高于母代经先高温后农药处理(HP)相应子代的 r_m,而在热持续3 d 下,母代先后胁迫次序处理对子代 r_m 的影响却与热持续1 d 和 5 d 情况相反[图3–11(A)]。此外,即使母代经历的是同一种高温农药处理,不同热持续天数也会对子代的 r_m 产生显著复杂

影响（$P<0.05$）。

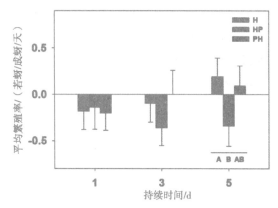

图 3-11 母代经历高温和吡虫啉农药不同胁迫处理后对子代成蚜繁殖率的影响（平均值 ± 标准误）。每个柱子表示处理组减去空白对照组对应的平均值与标准误。"*"表示处理组和空白对照组间差异达显著水平（$P=0.05$）。大写字母代表不同色柱处理方式间的差异显著性（$P=0.05$）。小写字母代表同色柱不同持续天数高温事件间的差异显著性（$P=0.05$）。无标记则为差异不显著。

Fig. 3-11 Effects of maternal exposing heat event interaction with imidacloprid in different treatments on offspring fecundity rate (mean ± SE) of S.avenae. Each bar represents the mean value and standard error corresponding to the different treatment minus the CK treatment. The "*" represent significant difference ($P=0.05$) between treatment and CK group. Different capital letters indicate significant level ($P=0.05$) of values between different treatments. Different lowercase letters indicate significant level ($P=0.05$) of values between heat wave with different durations. No mark means the difference is not significant.

3.2.3.7 子代种群参数

对于子代种群净增长率 R_0 而言，与空白对照所产子代的 R_0 相比，只有母代经 PH_1、HP_3 以及 PH_3 处理后，不同程度地刺激了子代的 R_0 外，母代经历其他处理均不同程度地降低了子代的 R_0（图 3-11B）。母代经高温单作和高温农药不同胁迫次序互作在热持续 1 d、3 d、5 d 均对子代的 R_0 产生了显著影响

（$P<0.05$），且母代经先农药后高温（PH）处理后子代的 R_0 均高于母代经先高温后农药处理（HP）相应子代的 R_0 [图 3-11(B)]。与此同时，和子代种群内禀增长率 r_m 影响相同，不同热持续天数同样会对子代的 R_0 产生显著复杂影响（$P<0.05$）。

对于子代平均世代时间 G 而言，与空白对照所产子代的 G 相比，只有母代经高温单作处理显著延长了子代的 G 外，高温和农药不同顺序互作均显著缩短了子代平均世代所需时间 [图 3-11(C)]。母代经高温单作和高温农药不同胁迫次序互作无论在热持续几天处理下均对子代的 G 产生了显著影响（$P<0.05$），且在热持续 1 d 和 5 d 下，母代经先农药后高温（PH）处理后子代的 G 均长于母代经先高温后农药处理（HP）相应子代的 G，而在热持续 3 d 下，母代先后胁迫次序处理对子代 G 的影响却与热持续 1 d 和 5 d 情况相反 [图 3-11(C)]。此外，与子代种群内禀增长率 r_m、净增长率 R_0 相同，不同热持续天数同样会对子代的 G 产生显著复杂影响（$P<0.05$）。

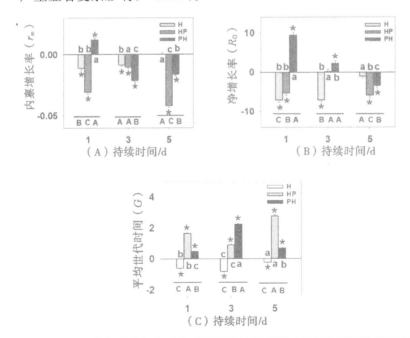

图 3-12 母代经历高温和吡虫啉不同胁迫处理后对子代种群参数的影响（平均值 ± 标准误）。每个柱子表示处理组减去空白对照组对应的平均值与标准误。"*"表示处理组和空白对照组间差异达显著水平（$P=0.05$）。大写

字母代表不同色柱处理方式间的差异显著性（P=0.05）。小写字母代表同色柱不同持续天数高温事件间的差异显著性（P=0.05）。

Fig. 3-12 Effects of maternal exposing heat event interaction with imidacloprid in different treatments on population parameters (mean ± SE) of *S.avenae*. Each bar represents the mean value and standard error corresponding to the different treatment minus the CK treatment. The "*" represent significant difference (P=0.05) between treatment and CK group. Different capital letters indicate significant level (P=0.05) of values between different treatments. Different lowercase letters indicate significant level (P=0.05) of values between heat wave with different durations.

◆ 3.3 结论与讨论

3.3.1 经历高温和吡虫啉不同胁迫次序处理后的当代生活史性状响应

高温和吡虫啉胁迫次序处理对麦长管蚜即时死亡率联合毒力结果不尽相同。例如，热持续 1 d，先高温后农药处理表现为拮抗作用，而先农药后高温处理却表现为协同作用；但在热持续 3 d 和 5 d 下，和农药先后不同顺序互作后却均表现为相加作用。从对即时死亡率的影响来看，短时的高温处理和农药互作后，会因为两胁迫处理顺序的不同产生不同的互作效应，但农药与中长时的高温先后不同顺序互作却均表现为相加作用。

通过对当代寿命、繁殖以及种群参数的调查发现，相较空白对照，无论是经历高温单作还是高温农药互作均会对当代麦长管蚜的寿命、繁殖、繁殖率以及种群参数产生显著负面影响，且相较高温单作，高温农药互作产生的负面作用更强烈，即高温农药互作后最终的联合毒力确实得到了提升，这与大多数胁迫互作后所得到的结果相同[181,182]。这可能由于生物在经历第一种胁迫后并没有完全恢复，所以当遭遇到第二种胁迫时，会对生物产生更加强烈的影响[183]。

此外,我们发现高温、吡虫啉不同胁迫次序处理对当代试虫寿命、两性繁殖性状产生的负面影响并不存在显著差异,但却对繁殖率和种群参数产生了不同的影响,这种影响因为高温持续天数的不同产生显著差异。

3.3.2 经历高温和吡虫啉不同胁迫次序处理后的跨代生活史性状响应

3.3.2.1 对子代若蚜发育和存活的影响

母代经历高温农药不同胁迫次序处理后不仅会影响试虫当代的生活史性状,而且这种影响甚至会延续到子代。母代无论经历高温单作还是高温农药互作均会延长子代发育所需时间、压低子代若虫存活率,这可能是由于母代经历胁迫后会将资源用于自身合成抵抗胁迫的物质[46]保证存活,从而减少子代资源的分配所致[184]。此外,母代经历高温农药先后胁迫次序处理也会对子代发育产生不同的后果,先高温后农药(HP)处理相比先农药后高温(PH)处理会延长子代的发育,对子代发育的负面影响更大(图3-6)。但两胁迫的胁迫次序处理对子代若虫存活率的影响确是相反的,先高温后农药处理相比先农药后高温对子代若虫存活产生负面影响更小。

3.3.2.2 对子代成蚜表现及种群参数的滞后效应

而母代经历胁迫后对子代成蚜产仔前期的影响,相对发育和存活而言相较复杂,既有正面的刺激也有负面的效应,但这些影响均比较微弱未达显著。且子代发育和成蚜产仔前期对于母代经历高温农药两胁迫次序的响应结果是一致的,均是先高温后农药处理相比先农药后高温处理产生的负面影响较大。

母代经历胁迫后对子代成蚜寿命和繁殖的影响一致,母代经历高温单作短期内(1 d或者3 d)可能会对子代的寿命、繁殖已经繁殖率产生负面影响,但长期(热持续5 d)就会产生微弱的正面影响,这可能是因为短期热处理对母代较弱的存活筛选作用所造成,如母代经历热1~3 d即时死亡率仅为0~26%,而长期

（5 d）热就会明显增加母代即死死亡率达72%,从而使得长时热存活下来的个体适合度相差不大,同时加上母代长时热对子代苷血存活的进一步筛选,更加使得留存下来的子代的寿命、繁殖和繁殖率得到了提升。而母代经历高温和农药互作无论是热持续短期还是长期均会不同程度的刺激子代寿命和繁殖,但对繁殖率的影响却差异不大。此外,子代成蚜寿命、繁殖、繁殖率对于母代经历高温农药两胁迫次序的响应结果是一致的,均是先农药后高温(PH)处理产生的正面刺激作用更大。

尽管母代经历两胁迫互作后,对子代的成蚜性状产生了正面刺激作用,但从种群参数来看,整体上母代经胁迫处理后均会对子代种群参数产生负面影响。与空白对照子代相比,除PH_1显著刺激子代r_m提升了6%以外,其余处理均显著压低了子代的r_m约4%~21%。此外,从我们的结果也可以看出:母代经历先农药后高温对子代种群参数整体的负面影响要低于先高温后农药的。

总之,麦长管蚜经高温农药先后不同胁迫次序处理后,会产生不同的即时死亡效应,虽对当代生活史性状(繁殖、寿命)的影响差异不显著,但会导致当代繁殖率、种群参数的不同。最重要的是,母代经高温农药先后不同胁迫次序处理后,也会产生不同的跨代效应,具体而言,先农药后高温(PH)对子代存活产生的负面影响更加强烈,但对子代的其他生活史性状产生的影响较小。本发现意味着,当生物暴露在多种胁迫环境时,各胁迫出现的次序也决定这该物种在当前环境下的种群命运。因此,本研究结果对气候变暖、农药广泛应用背景下麦蚜动态的准确预测及综合防控具有重要的指导意义。

第 4 章

第 4 章

热背景经历对麦长管蚜高温和农药胁迫响应的影响

环境背景经历在大多数生物的发育过程中都起着至关重要的作用[185]。生物的表型性状是基因和环境相互作用的结果[186],早期的环境干扰,可能会造成深刻的长期影响[187]。生物遭受恶劣的早期环境后,虽然可能会通过补偿效应在后期发育中得到恢复[188],但也可能会导致持久的表型变化。例如,个体早期的环境温度经历不但可以改变生物的热耐受性[138],影响其后的发育速率[189]、繁殖能力[41]等生活史轨迹;甚至可以产生跨代效应,影响后代的表现[184,190]。

此外,在多种胁迫环境下,热历史背景经历也可能会改变生物对其他胁迫的应对能力。例如,热带果蝇(*Zaprionus indianus*)经历热驯化会产生交互耐受性提升其应对干旱和饥饿胁迫的能力[191];类似的情况,在水生生物寡杜父鱼(*Oligocottus maculosus*)中也有报道,当该物种经历比饲养温度高12℃的高温热激后,会增加其对缺氧和高渗透胁迫的耐受性,而当热激高温提升至饲养温度高 15℃后,则会降低其应对随后所面临的缺氧和高盐胁迫的能力[192]。

为了明确不同热背景经历对生物响应高温和农药的影响,我们考察了麦长管蚜经历三种不同热背景后(22℃恒温,22℃+34℃ 2 h/d、22℃+38℃ 2 h/d),耐热性、农药敏感性以及当代和子代生活史性状、种群参数的变化。以期解决如下几个问题:

(1)热背景经历是否改变麦长管蚜耐热性,不同热背景经历影响是否不同?

(2)热背景经历是否改变麦长管蚜农药敏感性,不同热背景经历影响是否不同?

(3)麦长管蚜热背景经历和农药施用对当代及后代生活史性状及适合度是否产生显著延迟效应?

◆ 4.1 材料与方法

4.1.1 供试虫源

供试虫来源同第 1 章 1.1。

4.1.2 供试药剂

95% 吡虫啉原药(中农联合生物科技有限公司提供)。

4.1.3 试验因子设定

4.1.3.1 热背景设计

我们设置了三种热背景,即恒温 22℃为对照(NA)以及高低两个不同强度,最高温分别为 34℃(LA)和 38℃(HA)(图 4-1)。热背景处理模式以 24 h 为周期,在白天,温度从早上 8:00 的 22℃开始升温,到 12:00 达到最高分别为 34℃或 38℃,且在最高温维持 2 h,之后从下午 1:00 开始到下午 4:00 开始下降到 22℃,其余时间均保持在 22℃,以避免不同夜间温度所带来的影响[162]。这三种热背景分别有三台人工气候箱来控制,相对湿度在 50% 作用,光照为 G∶D=16∶8。

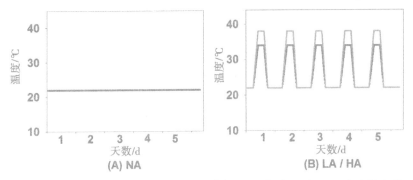

图 4-1 三种热背景经历设置。A:代表恒温 22℃(NA)。B:试验期内的高强度 HA(灰线)和低强度 LA(黑线)的热背景经历。

Fig.4-1 Three thermal acclimation conditions. NA(no acclimation), LA(low daytime temperature acclimation), and HA(high daytime temperature acclimation)represent a constant temperature of 22 ℃(A), 22℃ + 34℃ for 2 h per day(B:gray line), and 22℃ + 38℃ for 2 h per day(B:black line).

4.1.3.2 毒苗制备

制备过程同第 1 章 1.4.2。

4.1.4 试验设计

4.1.4.1 热背景经历对麦长管蚜耐热性(CT_{max})的影响

挑取饲养条件下的新生若蚜,分成 3 组分别放在不同的热历史背景,每组包括一个养虫笼,每笼放置一钵小麦苗,苗上接有 100 头新生若蚜。接入蚜虫时,小麦苗高度为 2~3 cm,为了避免蚜虫逃跑,中途不换麦苗。但定期浇水,以保证小麦苗的新鲜。根据预试验确定麦长管蚜在三种热背景 NA、LA、HA 下的若虫发育历期分别为 9.5(7~11)d,8.3(6~11)d 及 9.2(6~10)d。因此,为保证发育的整齐,所有的处理均在第 9 d 时,进行成蚜

CT_{max} 的测定。

CT_{max} 测定：参照 Zhao 等[138]的方法，将前处理后的蚜虫单独放在蜂巢板（80 mm × 80 mm）的小孔（直径 5 mm，深度 5 mm）中，一边粘贴上 200 目的尼龙纱网，另一边盖上透明玻璃板从而将蚜虫单头固定于各自的小孔中。之后将蜂巢板垂直放置在油浴的外接双层玻璃容器中，22℃下平衡 5 min 后，以 0.5℃/ min 速率升温至 30℃，再以较慢速率 0.1℃/ min 速率升温至 42℃，容器内的温度通过热电偶（Pt100）进行监控。用摄像机记录在此期间每头蚜虫触角或足最后一次移动时对应的温度（CT_{max}），每处理随机挑取蚜虫 30 头。

4.1.4.2 热背景经历对麦长管蚜响应农药胁迫生态表型影响

将 900 头新生若蚜，分成 6 组，150 头 / 组，分别进行各自的处理（图 4-1），具体如下。

温度处理：选取 3 组，单独分装于饲养管中，分别置于相应的热历史背景温度下。一旦若虫进入成蚜期，前处理结束，该蚜虫就转移到恒温 22℃，统一生存环境（图 4-1）。

温度 + 农药处理：将余下 3 组，同样单独分装于饲养管中，置于相应的热历史背景温度下。一旦若虫进入成蚜期，即将蚜虫接入插有毒苗的饲养管中置于恒温 22℃进行农药处理，处理 1 d 后，前处理结束。且将存活个体转移到新鲜无药的麦苗上，继续放置于 22℃，统一生存环境（图 4-1）。

母代性状调查：试验期间，所有处理每天上午 8:00 定期检查若虫发育以及存活并将已死若蚜移除，待若蚜发育至成蚜后，分别进行各自的处理，前处理结束后统计各自的死亡率。统一生存环境后，继续每天在 8:00 定期检查成蚜的存活以及产仔数目，直至供试蚜虫全部死亡，母代性状调查结束。成蚜寿命是指从变成成蚜到蚜虫死亡的时间。成蚜繁殖是指每头成蚜所产后代的总数。考虑到死亡率基数的一致性，因此我们仅调查 48 头蚜虫每处理的生活史性状，其余作为备用处理，仅用于子一代的采集。

跨代性状调查：前处理结束后第二天调查时（♀），随机采取 30 头子代 / 处理，置于常温 22℃养虫室内，同样于每日上午 8 点计数死亡个体、蜕皮及产仔情况,并将蜕皮、死亡蚜虫及新生若蚜去除,直至供试蚜虫全部死亡。去除逃逸蚜虫,所有处理蚜虫均测定了以下指标:若虫死亡率、发育历期、成蚜产仔前期、成蚜寿命以及繁殖。若虫死亡率是指活到成蚜的若虫占全部测试若虫的比例。成蚜寿命是指从变成成蚜到蚜虫死亡的时间。成蚜繁殖是指每头成蚜所产后代的总数。

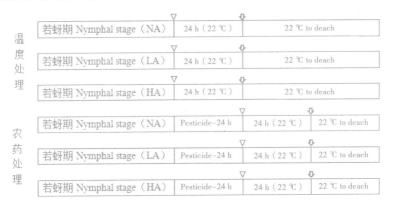

图 4-2 不同热背景下麦长管蚜响应农药吡虫啉胁迫处理流程图。"▽"代表前处理结束和生存率统计时刻。"⇩"代表子代取样点。NA 代表饲养条件下的热背景恒温 22℃,LA 和 HA 为试验期内的低强度和高强度热背景。

Fig. 4-2 Experimental design of S.avenae response to pesticide following different thermal acclimation conditions. indicate ending pretreatment and survival rate assessment. ⇩ indicate offsprings sampling points. NA: no acclimation; LA: low daytime temperature acclimation; HA: high daytime temperature acclimation.

4.1.5 统计分析

CT_{max} 以及母代寿命、繁殖性状均基本符合正态分布,因此,采用一般线性模型(GLM)进行方差分析,利用 Duncan 法进行热背景处理间多重比较,利用独立样本 t 检验来比较农药处理间的差异显著性(SPSS 19.0);子代生活史性状中若蚜发育历期和成蚜产仔前期,在分析之前,需经平方根转换后提高其正态性。转换后,子代若蚜发育历期、成蚜产仔前期、寿命、成蚜繁殖性状同样采用一般线性模型(GLM)进行方差分析,利用 Duncan 和独立

样本 t 检验来进行处理间差异显著性。虽然子代发育历期和产仔前期性状均经过了数据转换,但作图为了直观仍用原始数据。子代生存曲线的绘制和差异显著性比较分别采用 Kaplan-Meier 法和 Log Rank 法。所有存活率显著性分析均采用列联表法进行分析,多重比较通过非独立 2×2 表法实现。

种群参数均根据实际记录计算。计算和分析同第 1 章 1.6 种群参数的统计分析。

◆ 4.2 结果与分析

4.2.1 热背景经历对麦长管蚜耐热性的影响

不同热背景经历对麦长管蚜的耐热性产生显著影响($F_{2,84}$= 30.653,$P<0.001$),且随着热历史强度的增加表现出同步升高的趋势(图 4-3)。与恒温对照 22℃下麦长管蚜耐热性(38.54±0.99)℃相比,经历低强度的热背景处理后,耐热性增加了(1.17±0.17)℃,且随着热背景强度的进一步提升,耐热性同步提升了(1.25±0.18)℃(图 4-3)。此外,麦长管蚜经高低强度热背景处理后,耐热性相差幅度很小,且并未表现出差异显著性(图 4-3)。

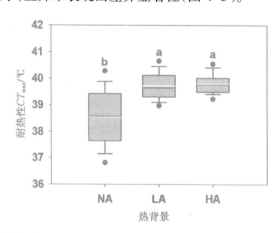

图 4-3 麦长管蚜经历不同热背景后,耐热性 CT_{max} 的动态变化。小写字母代表不同热背景处理间的差异显著性($P=0.05$)。箱中的白线和黑线分别

代表平均值和中值。黑点表示离群值。

Fig. 4-3 Effect of thermal acclimation condition on CT_{max} in the S.avenae. Different letters above the plots indicate significant differences between acclimation conditions (P=0.05).The white and black lines inside the boxes represent the mean and median values, respectively. Black dots mark the outliers (only showing 5th/95th).

4.2.2 热背景经历对当代成蚜响应农药胁迫的影响

4.2.2.1 存活

麦长管蚜经历不同热背景后,对存活产生了显著影响(图4-4浅灰柱,$\chi^2 = 49.192$, $df = 2$, $P <0.001$)。与对照存活率95.8%相比,随着热背景强度的提升,存活率表现出下降的趋势,分别是低强度热背景(LA)下的87.5%以及高强度热背景(HA)下的37.5%(图4-4浅灰柱)。同样的,不同热背景也会对麦长管蚜响应农药胁迫产生显著影响(图4-4深灰柱,$\chi^2 = 41.255$, $df = 2$, $P <0.001$)。与对照热背景下农药所导致的存活率64.6%相比,经历低强度热背景处理后,增强了试虫对农药的抵抗力提高了试虫的存活至79.2%,但高强度的热背景,增加了试虫对农药的敏感性进一步降低了试虫的存活率至16.7%(图4-4深灰柱)。此外,无论麦长管蚜经历何种热背景,农药的存在均会降低试虫的存活,且在热背景处理 NA($\chi^2 = 14.764$, $df = 1$, $P <0.001$)和 HA($\chi^2 = 5.275$, $df = 1$, $P = 0.022$)下达到了显著水平(图4-4)。

4.2.2.2 寿命

麦长管蚜经历不同热背景后,对寿命产生了显著影响(图4-5浅灰箱,$F_{2,103}= 15.135$, $P <0.001$)。与对照成蚜寿命(17.5 ± 7.58)d相比,随着热=背景强度的提升,寿命表现出越来越明显的下降趋势,分别是低强度热背景(LA)下的(16.9 ± 7.06)d以及高强度热背景(HA)下的(7.67 ± 1.57)d(图4-5浅灰箱)。同样的,麦长管蚜经历不同热背景后施药对其寿命也会产生显著影

响(图4-5 深灰箱,$F_{2,76}= 8.893$, $P <0.001$)。与对照热背景施加农药后寿命(10.1 ± 7.01)d 相比,经历低强度热背景处理后,农药的施加反而刺激了寿命延长至(15.6 ± 7.45)d,但高强度的热背景,反而再一次压低了试虫的寿命至(6.56 ± 3.75)d(图4-5 深灰箱)。此外,农药的施用在试虫经对照 CK($t = 4.357$, $df = 76$, $P<0.001$)背景处理后显著地缩短了成蚜的寿命,但经热背景 LA($t = 0.815$, $df = 78$, $P = 0.417$)和 HA($t = 0.853$, $df = 9.437$, $P = 0.415$)处理后,农药的施用对麦长管蚜的寿命无显著影响(图4-5)。

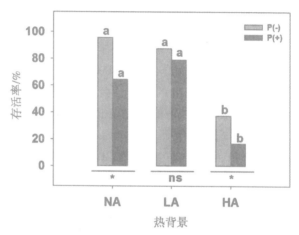

图4-4 麦长管蚜经历不同热背景后,农药的施用对麦长管蚜存活的影响。小写字母代表同色柱不同热背景处理间的差异显著性($P=0.05$)。"*"和"ns"分别表示相邻柱间差异达显著或不显著水平($P=0.05$)。

Fig. 4-4 Effect of pesticide on survival rate in the *S.avenae* following different thermal acclimation condition. Different letters above the bars indicate significant differences between acclimation conditions ($P=0.05$). "*" and "ns" represent significant and nonsignificant differences between the pesticide present P(+) and absent P(-) treatment, respectively.

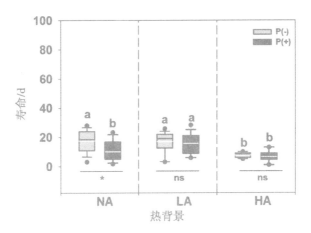

图 4-5 麦长管蚜经历不同热背景后,农药的施用对麦长管蚜寿命的影响。小写字母代表同色箱图不同热背景处理间的差异显著性($P=0.05$)。"*"和"ns"分别表示相邻两箱间差异达显著或不显著水平($P=0.05$)。箱中的白线和黑线分别代表平均值和中值。黑点表示离群值。

Fig. 4-5 Effect of pesticide on longevity in the *S. avenae* following different thermal acc limation condition. Different letters above the plots indicate significant differences between acclimation conditions ($P=0.05$). "*" and "ns" represent significant and nonsignifcant differences between the pesticide present P (+) and absent P (-) treatment, respectively. The white and black lines inside the boxes represent the mean and median values, respectively. Black dots mark the outliers.

4.2.2.3 繁殖

麦长管蚜经历不同热背景处理后农药的有无对繁殖的影响与寿命相类似。经历不同热背景处理后,对麦长管蚜繁殖产生了显著影响(图 4-6 浅灰箱, $F_{2,99}= 26.195$, $P <0.001$)。与对照成蚜繁殖力(37.2 ± 15.5)头若蚜/成蚜相比,随着热历史背景强度的提升,繁殖表现出下降趋势,分别是低强度热背景(LA)下的(28.5 ± 17.5)头若蚜/成蚜以及高强度热背景(HA)下的(6.4 ± 1.9)头若蚜/成蚜(图 4-6 浅灰箱)。同样的,麦长管蚜经历不同热背景后施药对其繁殖也会产生显著影响(图 4-6 深灰箱, $F_{2,74}= 5.053$, $P =0.009$)。与对照热背景施加农药后繁殖力

（18.0±14.7）头若蚜/成蚜相比，经历低强度热背景处理后，农药的施加反而刺激了繁殖力到（23.1±13.0）头若蚜/成蚜，但高强度的热背景，反而再一次压低了试虫的繁殖力至（7.3±1.5）头若蚜/成蚜（图4-6深灰箱）。此外，农药的施用在试虫经对照NA（$t = 5.332$, $df = 71$, $P<0.001$）背景处理后显著的压低了成蚜的繁殖，但经热背景LA（$t = 1.565$, $df = 78$, $P=0.122$）和HA（$t = -0.552$, $df = 8.216$, $P=0.595$）处理后，农药的施用对麦长管蚜的繁殖无显著影响（图4-6）。

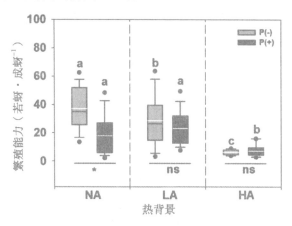

图4-6 麦长管蚜经历不同热背景后，农药的施用对麦长管蚜繁殖的影响。小写字母代表同色箱图不同热背景处理间的差异显著性（$P=0.05$）。"*"和"ns"分别表示相邻两箱间差异达显著或不显著水平（$P=0.05$）。箱中的白线和黑线分别代表平均值和中值。黑点表示离群值。

Fig. 4-6 Effect of pesticide on fecundity in the *S.avenae* following different thermal acclimation condition. Different letters above the plots indicate significant differences between acclimation conditions（$P=0.05$）. "*" and "ns" represent significant and nonsignifcant differences between the pesticide present P（+）and absent P（-）treatment, respectively. The white and black lines inside the boxes represent the mean and median values, respectively. Black dots mark the outliers（only showing 5th/95th）.

4.2.2.4 繁殖率

麦长管蚜经历不同热背景处理后农药的有无对繁殖率的影

响与寿命、繁殖相类似。经历不同热背景处理后,对麦长管蚜繁殖率产生了显著影响(图4-7浅灰箱,$F_{2,99}$= 27.717, P <0.001)。与对照成蚜每天繁殖率(2.0±0.6)头若蚜/成蚜相比,随着热背景强度的提升,每天繁殖率表现出下降趋势分别是低强度热背景(LA)下的(1.6±0.6)头若蚜/成蚜以及高强度热背景(HA)下的(0.8±0.2)头若蚜/成蚜(图4-7浅灰箱)。同样的,麦长管蚜经历不同热背景后施药对其繁殖率也会产生显著影响(图4-7深灰箱,$F_{2,74}$= 5.483, P =0.006)。与对照热背景施加农药后每天繁殖率(1.6±0.6)头若蚜/成蚜相比,经历低强度热背景处理后,农药的施用使得每天繁殖率稍有下降为(1.5±0.4)头若蚜/成蚜/天,但高强度的热背景,农药的施用使得每天繁殖率下降显著为(1.0±0.2)头若蚜/成蚜(图4-7深灰箱)。此外,农药的施用在试虫经对照NA(t = 2.674, df = 71, P=0.009)背景处理后显著的压低了成蚜的繁殖率,但经热背景LA(t = 1.050, df = 78, P=0.297)和HA(t = −1.345, df = 24, P=1.191)处理后,农药的施用对麦长管蚜的繁殖率无显著影响(图4-7)。

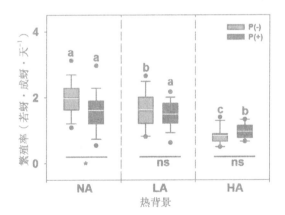

图4-7 麦长管蚜经历不同热背景后,农药的施用对麦长管蚜繁殖率的影响。小写字母代表同色箱图不同热背景处理间的差异显著性(P=0.05)。"*"和"ns"分别表示相邻两箱间差异达显著或不显著水平(P=0.05)。箱中的白线和黑线分别代表平均值和中值。黑点表示离群值。

Fig. 4-7 Effect of pesticide on fecundity rate in the *S.avenae* following different thermal acclimation condition. Different letters above the plots indicate significant differences between acclimation conditions ($P=0.05$). "*" and "ns" represent significant and nonsignifcant differences between the pesticide present P (+) and absent P (–) treatment, respectively. The white and black lines inside the boxes represent the mean and median values, respectively. Black dots mark the outliers (only showing 5th/95th).

4.2.2.5 种群参数

r_m,内禀增长率; R_0,净繁殖率; G,平均世代周期。N 和 Y 分别代表农药的无和有。

生命表参数结果表明,随着热背景强度的增加,种群内禀增长率(r_m)(图 4-8)、净繁殖率(R_0)以及平均世代周期(G)均呈现越来越明显的下降趋势(表 4-1)。而当经历不同热背景后施药,种群内禀增长率(r_m)仍然随着热背景强度的增加呈现越来越明显的下降趋势(图 4-8),但净繁殖率(R_0)和平均世代周期(G)却在经历低强度的热背景后得到了提升,经历高强度的热背景处理后两性状的表现下降(表 4-1)。此外,无论麦长管蚜前期经历哪个热背景经历,农药的施用均会对试虫的种群参数产生显著的负面影响(表 4-1,图 4-8)。

表 4–1 麦长管蚜经历不同热背景后,农药的施用对麦长管蚜种群参数的影响

Table 4–1 Effect of pesticide on population parameters in the *S.avenae* following different thermal acclimation condition.

热处理 Heat treatment	农药 Pesticide	种群参数(平均值 ± 标准误)		
		r_m	R_0	G
NA	N	0.19 ± 0.00	32.42 ± 0.27	18.53 ± 0.03
	Y	0.18 ± 0.00	11.28 ± 0.21	13.72 ± 0.04
LA	N	0.19 ± 0.00	24.94 ± 0.29	17.05 ± 0.03
	Y	0.18 ± 0.00	18.21 ± 0.2	15.81 ± 0.05
HA	N	0.07 ± 0.00	2.36 ± 0.05	12.45 ± 0.03
	Y	0.01 ± 0.00	1.16 ± 0.04	13.54 ± 0.06

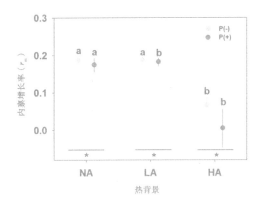

图 4-8 麦长管蚜经历不同热背景后,农药的施用对麦长管蚜种群内禀增长率(平均值 ±95% 置信区间)的影响。小写字母代表同色散点不同热背景处理间的差异显著性($P=0.05$)。"*"表示相邻两点间差异达显著水平($P=0.05$)。

Fig. 4-8 Effect of pesticide on intrinsic rates of population increase (mean ± 95%CI) in the *S.avenae* following differernt thermal acclimation condition. Different letters indicate significant differences between acclimation conditions ($P=0.05$). "*" represent significant differences between the pesticide present P(+) and absent P(−) treatment.

4.2.3 热背景和吡虫啉经历对跨代生活史性状的影响

4.2.3.1 子代若蚜发育历期

母代单独经不同热背景处理后,对子代若蚜发育历期产生了显著影响(图 4-9 浅灰箱,$F_{2,80}= 19.400, P <0.001$)。与对照子代发育历期($7.67 ± 0.55$)d 相比,母代经历低强度的热背景处理后对子代发育历期的影响较弱为($7.53 ± 0.57$)d,而母代高强度热背景经历却显著延长了子代发育历期至($8.65 ± 0.98$)d(图 4-9 浅灰箱)。母代若蚜期经历不同热背景处理后,成蚜期再经农药处理对子代的若蚜历期也会显著影响(图 4-9 深灰箱,$F_{2,72}= 52.377, P <0.001$)。与对照背景下施药子代发育历期($7.46 ± 0.59 R_o$)d 相比,随着母代经历热背景强度的提升,子代

发育历期表现明显延长的趋势,分别是低强度热背景(LA)下的(7.92±0.87)d以及高强度热背景(HA)下的(9.71±0.91)d(图4-9深灰箱)。此外,母代农药处理,在对照CK(t=4.357, df=76, $P<0.001$)背景下对子代发育历期无显著影响,但在低强度LA(t=-2.017, df=55, P=0.049)和高强度HA(t=-3.947, df=48, $P<0.001$)背景下却显著延长了子代的发育历期(图4-9)。

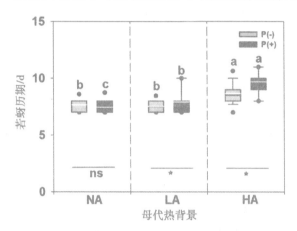

图4-9 母代不同热背景和农药处理经历对子代若蚜历期的影响。小写字母代表同色箱不同热背景处理间的差异显著性(P=0.05)。"*"和"ns"分别表示相邻两箱间差异达显著或不显著水平(P=0.05)。箱中的白线和黑线分别代表平均值和中值。黑点表示离群值。

Fig. 4-9 Effects of maternal exposing different acclimation conditions interaction with pesticide on offspring nymphal time of $S. avenae$. Different letters above the plots indicate significant differences between acclimation conditions (P=0.05). "*" and "ns" represent significant and nonsignifcant differences between the present P(+) and absent P(-) treatment, respectively. The white and black lines inside the boxes represent the mean and median values, respectively. Black dots mark the outliers.

4.2.3.2 子代若蚜存活率

无论母代是否经农药处理,母代不同热背景经历对子代若蚜存活产生的影响趋势一致,但均未达显著水平(图4-10浅灰柱,

$\chi^2 = 4.243$, $df = 2$, $P = 0.120$；图 4-10 深灰柱，$\chi^2 = 1.440$, $df = 2$, $P = 0.487$）。与对照背景下的若蚜存活率相比 [P(-)：93%；P(+)：80%]，母代经低强度的 LA 热背景处理后，子代若蚜存活率会出现小幅的提升 [P(-)：100%；P(+)：90%]，而母代经高强度的 HA 热背景处理后，子代若蚜存活率会下降 [P(-)：87%；P(+)：79%]。此外，母代农药处理，在所有热背景下均没有对子代若蚜存活率产生显著影响（图 4-10，CK：$\chi^2 = 2.160$, $df = 1$, $P = 0.142$；LA：$\chi^2 = 3.158$, $df = 1$, $P = 0.076$；HA：$\chi^2 = 0.480$, $df = 1$, $P = 0.488$）。

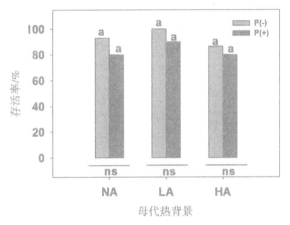

图 4-10 母代经历不同热背景和农药处理后对子代若蚜存活率的影响。小写字母代表同色柱不同热背景处理间的差异显著性（$P=0.05$）。"ns" 表示相邻柱间差异未达显著水平（$P=0.05$）。

Fig. 4-10 Effects of maternal exposing different acclimation conditions interaction with pesticide on offspring survival rate of *S. avenae*. Different letters above the bars indicate significant differences between acclimation conditions（$P=0.05$）. "ns" represent nonsignificant differences between the pesticide present P（+）and absent P（-）treatment.

4.2.3.3 子代成蚜产仔前期

母代单独经不同热背景处理后，对子代成蚜产仔前期产生了显著影响（图 4-11 浅灰箱，$F_{2,79}= 5.729$, $P =0.005$）。与对照子代发育历期（0.35 ± 0.49）d 相比，母代经历低强度的热背景处理后对子代成蚜产仔前期的影响较弱为（0.20 ± 0.41）d，而母代高强度热背景经历却显著延长了子代成蚜产仔前期至(0.62 ± 0.50)d（图4-11浅灰箱）。母代经历不同热背景处理后，再经农药处理对子代成蚜产仔前期无显著影响（图4-11深灰箱，$F_{2,72}= 0.226$，$P =0.799$）。此外，母代农药处理，在所有热背景下均没有对子代成蚜产仔前期产生显著影响（图4-11，NA：$t = 0.094$, $df = 48$, $P=0.926$；LA：$t = -1.776$, $df = 48.878$, $P=0.082$；HA：$t = 1.855$, $df = 48$, $P=0.070$）。

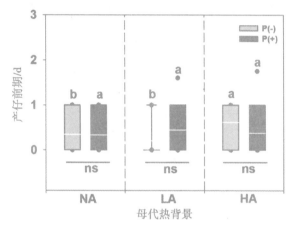

图4-11 母代经历不同热背景和农药处理后对子代成蚜产仔前期的影响。小写字母代表同色箱图不同热背景处理间的差异显著性（$P=0.05$）。"ns"表示相邻柱间差异未达显著水平（$P=0.05$）。箱中的白线和黑线分别代表平均值和中值。黑点表示离群值。

Fig. 4-11 Effects of maternal exposing different acclimation conditions interaction with pesticide on offspring pre-productive period of *S.avenae*. Different letters above the plots indicate significant differences between acclimation conditions ($P=0.05$). "ns" represent nonsignifcant differences between the pesticide present P（+）and absent P（-）treatment. The white and black lines inside the boxes represent the mean and median values, respectively. Black dots mark the outliers.

4.2.3.4 子代成蚜寿命

母代单独经不同热背景处理后，对子代成蚜寿命产生了显著影响且随着经历热历史背景强度的提升，对子代的寿命产生了刺激作用（图 4-12 浅灰箱，$F_{2,80}$= 3.408，P = 0.038）。与对照子代成蚜寿命（15.41 ± 8.60）d 相比，母代经历低强度和高强度的热背景处理后子代寿命分别延长至（18.13 ± 7.11）d 和（20.92 ± 7.32）d（图 4-12 浅灰箱）。母代经历不同热背景处理后，再经农药处理对子代的成蚜寿命也会显著影响，且同样随着经历热历史背景强度的提升，对子代的寿命产生了刺激作用（图 4-12 深灰箱，$F_{2,72}$= 3.797，P = 0.027）。与对照背景下施药子代寿命（17.42 ± 6.77）d 相比，母代经历低强度和高强度的热背景处理后子代寿命分别延长至（17.48 ± 5.84）d 和（21.83 ± 6.69）d（图 4-12 深灰箱）。此外，母代农药处理，在所有热背景处理下对子代成蚜寿命均未产生显著影响（图 4-12，NA：t = −0.918，df = 49，P=0.363；LA：t = 0.376，df = 55，P= 0.709；HA：t = −0.458，df = 48，P= 0.649）。

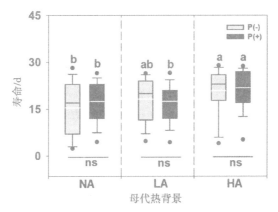

图 4-12 母代经历不同热背景和农药处理后对子代成蚜寿命的影响。小写字母代表同色箱图不同热背景处理间的差异显著性（P=0.05）。"ns"表示相邻柱间差异未达显著水平（P=0.05）。箱中的白线和黑线分别代表平均值和中值。黑点表示离群值。

Fig. 4-12 Effects of maternal exposing different acclimation conditions interaction with pesticide on offspring longevity of *S. avenae*. Different letters above the plots indicate significant differences between acclimation conditions（P=0.05）. "ns" represent nonsignifcant differences between the pesticide present P（+）and absent P（−）treatment. The white and black lines inside the boxes represent the mean and median values, respectively. Black dots mark the outliers.

4.2.3.5 子代成蚜繁殖

母代经历不同热背景和农药处理后对子代成蚜繁殖的影响与对子代成蚜寿命的影响相类似。母代单独经不同热背景处理后，对子代成蚜繁殖产生了显著影响且随着经历热历史背景强度的提升，对子代的繁殖产生了刺激作用（图 4-13 浅灰箱，$F_{2,79}$= 3.775, P = 0.027）。与对照子代成蚜繁殖力（37.73 ± 20.02）头若蚜/成蚜相比，母代经历低强度和高强度的热背景处理后子代繁殖力分别提升至（46.27 ± 19.01）头若蚜/成蚜和（52.00 ± 17.41）头若蚜/成蚜（图 4-13 浅灰箱）。母代经历不同热背景处理后，再经农药处理对子代的成蚜繁殖虽未产生显著影响（图 4-13 深灰箱，$F_{2,72}$= 1.994, P = 0.144），但仍然随着经历热历史背景强度的提升，对子代的繁殖产生了刺激作用。与对照背景下施药子代繁殖力（45.29 ± 17.22）头若蚜/成蚜相比，母代经历低强度和高强度的热背景处理后子代繁殖力分别提升至（45.48 ± 14.18）头若蚜/成蚜和（53.04 ± 15.04）头若蚜/成蚜（图 4-13 深灰箱）。此外，母代农药处理，在所有热背景处理下对子代成蚜繁殖均未产生显著影响（图 4-13，NA：t = −1.426, df = 48, P=0.160；LA：t = 0.175, df = 55, P= 0.862；HA：t = −0.225, df = 48, P= 0.823）。

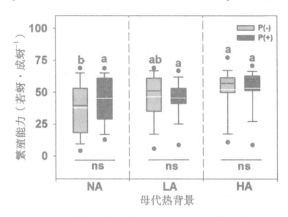

图 4-13 母代经历不同热背景和农药处理后对子代成蚜繁殖的影响。小写字母代表同色箱图不同热背景处理间的差异显著性（P=0.05）。"ns"表示相邻柱间差异未达显著水平（P=0.05）。箱中的白线和黑线分别代表平均值和中值。黑点表示离群值。

Fig. 4-13 Effects of maternal exposing different acclimation conditions interaction with pesticide on offspring fecundity of S.avenae. Different letters above the plots indicate significant differences between acclimation conditions ($P=0.05$). "ns" represent nonsignifcant differences between the pesticide present P (+) and absent P (−) treatment. The white and black lines inside the boxes represent the mean and median values, respectively. Black dots mark the outliers.

4.2.3.6 子代成蚜繁殖率

母代经历不同热背景和农药处理后对子代成蚜繁殖率的两因子方差分析结果表明：母代单独经历不同热背景处理（$F_{2,151}= 0.26, P = 0.77$）以及农药处理（$F_{2,151}= 1.25, P = 0.27$）对子代繁殖率均没有产生显著影响，且两者对子代该性状也不存在交互作用（$F_{2,151}= 0.90, P = 0.41$）（表 4-2）。

表 4-2 母代经历不同热背景和农药处理后对子代成蚜繁殖率的方差分析

Table 4-2 Results of variance analysis for effects of maternal exposing different acclimation conditions interaction with pesticide on offspring fecundity rate of S.avenae

母代处理 Maternal treatment	DF	MS	F	P
若蚜期热处理 Heat treatment of Nymph（HTN）	2	0.09	0.26	0.77
成蚜期农药处理 Pesticide treatment of adult（ITA）	1	0.42	1.25	0.27
若蚜期热处理 × 成蚜农药处理 HTN × ITA	2	0.3	0.90	0.41
误差 Error	151	0.33	—	—

4.2.3.7 子代生存曲线

无论母代是否经农药处理，母代不同热背景经历对子代生存分布产生的影响趋势一致，且均达显著水平 [图 4-14(A), $\chi^2 = 4.243, df = 2, P = 0.120$；图 4-14(B), $\chi^2 = 15.138, df = 2, P <0.001$]。与对照背景下子代生存的中位时间 [P(−): 22；P(+): 23 d] 相比，随着母代经历热背景强度的提升，对子代生存中位时

间的延长越来越明显,具体而言,母代经低强度的 LA 热背景处理后,子代生存的中位时间会得到延长分别至 P(–): 28 d 和 P(+): 27 d,而母代经高强度的 HA 热背景处理后,子代生存的中位时间会得到进一步延长分别至 P(–): 31 d 和 P(+): 31 d。此外,母代农药处理,在所有热背景下均没有对子代生存分布产生显著影响(图 4–14,NA: $\chi^2 = 0.600$, $df = 1$, $P = 0.439$; LA: $\chi^2 = 1.373$, $df = 1$, $P = 0.241$; HA: $\chi^2 = 1.664$, $df = 1$, $P = 0.197$)。

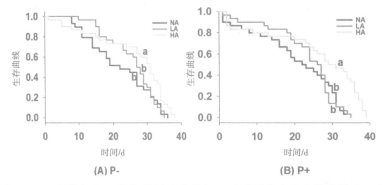

图 4–14 母代经历不同热背景和农药处理后对子代生存曲线的影响。小写字母代表不同热背景处理间的差异显著性($P=0.05$)。(A)母代单独热背景经历对子代生存曲线的影响(B)不同热背景经历后施加农药对子代生存曲线的影响。

Fig. 4–14 Effects of maternal exposing different acclimation conditions interaction with pesticide on offspring survival curve of *S.avenae*. Different letters above the plots indicate significant differences between acclimation conditions ($P=0.05$). (A) represent pesticide absent, P(–). (B) represent pesticide present, P(+).

4.2.3.8 子代种群参数

总体上看,无论母代是否经历农药处理,母代低强度的热背景经历,提高了子代种群参数的表现,而母代高强度的热背景经历,却对子代种群参数产生了负面影响。此外,母代农药处理对子代种群参数的刺激增长作用仅在对照背景下表现出来,而在母代经不同强度的热背景处理后,这种对子代种群参数的刺激作用就消失了(表 4–3)。以种群内禀增长率 r_m 为例:与对照背景下子代相比,母代经低强度的热背景处理后 r_m 整体提升了 1.6% P(-) 和 11.5% P(+),而母代经高强度的热背景处理后又分别将 r_m 压低了 1.7% P(-) 和 10.9% P(+)。此外,母代农药处理在 CK 背景下会刺激子代 r_m 整体提升 3.3%,但在低强度的热背景和高强度的热背景下则分别压低了 5.9% 和 6.4%(图 4-15)。

表 4-3 母代经历不同热背景和农药处理后对子代种群参数的影响

Table 4-3 Effects of maternal exposing different acclimation conditions interaction with pesticide on offspring population parameters of *S. avenae*.

热处理 Maternal condition	母代农药 Maternal pesticide	种群参数(平均值 ± 标准误)		
		r_m	R_0	G
NA	N	0.23 ± 0.00	34.09 ± 0.37	15.25 ± 0.04
	Y	0.24 ± 0.00	36.27 ± 0.46	15.02 ± 0.03
LA	N	0.26 ± 0.00	46.03 ± 0.33	14.86 ± 0.03
	Y	0.24 ± 0.00	40.55 ± 0.31	15.27 ± 0.03
HA	N	0.23 ± 0.00	45.23 ± 0.42	16.77 ± 0.03
	Y	0.21 ± 0.00	42.64 ± 0.50	17.63 ± 0.03

注:r_m,内禀增长率;R_0,净繁殖率;G,平均世代周期。N 和 Y 分别代表农药的无和有。

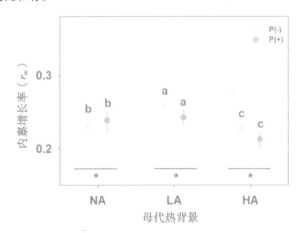

图 4-15 母代经历不同热背景和农药处理后对子代种群内禀增长率(平均值 ±95% 置信区间)的影响。小写字母代表同色散点不同热背景处理间的差异显著性(P=0.05)。"*"表示相邻两点间差异达显著水平(P=0.05)。

Fig. 4-15 Effects of maternal exposing different acclimation conditions interaction with pesticide on offspring intrinsic rates of population increase of S. avenae. Different letters indicate significant differences between acclimation conditions ($P=0.05$). "*" represent significant differences between the pesticide present P (+) and absent P (-) treatment.

◆ 4.3 结论与讨论

4.3.1 热背景经历对当代后续耐热性及生活史性状的影响

麦长管蚜若蚜期经历不同热背景后成蚜 CT_{max} 的测定结果表明：经历不同强度的热处理后，成蚜耐热性相较基底耐热性得到显著提升，且随着热背景强度增加耐热性出现同步增长的趋势。麦长管蚜所表现出的耐热性升高的这种适应性响应非常普遍，几乎发现于所有的生物，如节肢动物[33,193]、软体动物[194,195]、爬行动物[196]甚至鸟类[197]。事实上，生物的这种耐热性提高的适应性变化，是保证存活下来的关键，而保证存活同样是种群发展的首要前提条件。

通过对当代存活、寿命、繁殖、繁殖率以及种群参数的调查发现，相较对照背景，低强度的热背景驯化不仅提高了成蚜的耐热性，而且同时也未对上述生活史性状产生显著的负面影响。说明低强度的热背景驯化在提高耐热性的同时也未引起适合度代价的产生，该发现与适合性响应常伴随适合度代价这一普遍结论相反[198,199]。这一现象的产生可能涉及多种原因，热诱导的毒物兴奋效应，也即在经历了较低的热胁迫后，成虫相关性状（寿命、繁殖、耐热性）会得到提升[200,201]；补偿效应，通常被定义为在生物经历胁迫环境后，被转移到适宜环境通过快速的补偿生长得到恢复，会中和之前胁迫所带来的负面影响[166,167]；模块效应，可以将早期不利的生活环境隔绝，避免对之后的生活阶段产生影响，从而减少适合度代价的产生[168,169]。而高强度的热背景驯化虽然提高了成蚜的耐热性，但却对当代存活、繁殖、寿命以及种群参数产生了显著的负面影响。这可能与驯化的强度有关系,该驯化强度对于试虫而言可能已经非常接近其基底致死温度,虽然经历该强度背景的驯化后一定程度上提高了存活个体的热耐受性,但更大程度上,导致了一些无法修复的热损伤[156,202],从而使得后续的适合度性状大幅度下降。

4.3.2 热背景经历对当代农药胁迫生活史性状响应的影响

此外,母代经历不同热背景处理后,再经农药处理,相较对照背景下施药对当代存活、寿命、繁殖、繁殖率以及种群参数的影响,低强度的热背景驯化不仅未对母代上述适合度性状产生负面影响,甚至还产生了轻微的刺激作用。例如,对照热背景下施药后存活率为64.6%,而低强度热背景经历后施药存活率为79.2%;对照热背景下施药后当代寿命、繁殖分别为(10.1±7.01)和(18.0±14.7)头若蚜/成蚜,而低强度热背景经历后施药当代寿命、繁殖分别为(15.6±7.45)头若蚜/成蚜和(23.1±13.0)头若蚜/成蚜。但高强度的热背景驯化后施药与对照背景下施药相比,对当代存活、寿命、繁殖、繁殖率以及种群参数却产生了显著的负面影响。最重要的是,农药的施用在对照背景后,对当代生活史性状均产生了显著的负面影响,明显压低了存活、寿命、繁殖、繁殖率以及种群参数,但在低强度的背景下,农药施用对当代生活史性状产生影响较小。这说明低强度热背景对当代生活史性状所产生的正面刺激作用抵消了农药在该背景下所产生的负面影响。也可能低强度热背景经历诱导产生了交互耐受性,从而增强了生物对农药胁迫的耐受能力,这种不同胁迫类型互作诱导产生交互耐受性的报道非常普遍[191,192]。但是在高强度的热背景经历后施药,仅对当代的存活产生了显著的负面作用,而对当代寿命和繁殖却并没有产生进一步的负面影响,但却压低了当代整体的种群参数,这说明了生物在经历高强度的热背景驯化后,反而降低了生物对于农药的抵抗能力,由于两胁迫互作后对存活产生了更强的汰选作用,因此对后续的寿命和繁殖生活史性状产生的影响较小。

4.3.3 热背景和农药经历的跨代生活史性状响应

4.3.3.1 对子代发育和存活的影响

无论母代是否经历农药处理,母代热背景处理对子代发育历期、成蚜产仔前期的影响相同。基本为高强度热背景处理后发育所需时间最长,而对照背景处理和低强度热背景处理较差。这说明高强度的热背景处理不仅对当代性状产生了负面影响,且这种影响甚至会延续到子代,延长子代发育所需的时间。产生这一后果可能是由于母代的资源以及能量多用来合成耐热性物质[203]保证存活,从而减少了子代资源的分配以致延长发育所需的时间[204]。此外,与未施药的处理相比,母代农药处理仅在高低强度的热背景经历下会显著延长若虫的发育,而在对照背景下却不会产生母代效应,说明生物有限的应对能力只能应对农药这种单一胁迫,当热处理和农药胁迫同时存在时,就会使得胁迫所产生的负面影响传递到子代。

若虫存活率结果表明:无论母代是否经历农药处理,母代不同热背景经历对子代的若虫存活影响较小,且母代农药经历在所有的热背景下,同样不会对子代若虫存活产生影响。这说明,母代经历胁迫处理后对子代的负面影响仅延长了发育所需时间,且这一负面影响在这期间得到了修复,因此并未对存活产生负面影响。

4.3.3.2 对子代成蚜表现、生存分布及种群参数的滞后效应

与子代发育和存活不同,子代成蚜寿命、繁殖以及生存分布结果表明:无论母代是否经历农药处理,母代热背景处理对子代成蚜寿命、繁殖以及生存分布产生了正面刺激作用,且随着热背景强度的逐渐增加,产生的正面刺激作用逐渐显著。例如,对照子代成蚜寿命 P(-):(15.41 ± 8.60)d、P(+):(17.42 ± 6.77)d,母代经历低强度和高强度的热背景处理后子代寿命分别延长至 P(-):(18.13 ± 7.11)d、P(+):(17.48 ± 5.84)d 和 P(-):(20.92 ± 7.32)d、

P(+):(21.83±6.69)d。

对照子代成蚜繁殖 P(-):(37.73±20.02)头若蚜/成蚜、P(+):(45.29±17.22)头若蚜/成蚜,母代经历低强度和高强度的热背景处理后子代繁殖量分别增加至 P(-):(46.27±19.01)头若蚜/成蚜、P(+):(45.48±14.18)头若蚜/成蚜和 P(-):(52.00±17.41)头若蚜/成蚜、P(+):(53.04±15.04)头若蚜/成蚜。对照子代生存中位时间 P(-):(22 d、P+:23)d,母代经历低强度和高强度的热背景处理后子代生存中位时间分别延长至 P(-):28 d,P(+):27 d 和 P(-):31 d,P(+):31 d。这更加说明母代高强度热背景经历后所产生的负面影响仅延续到了子代发育,这种负面影响在对子代若虫存活性状的影响得到了缓解,延续到子代成蚜寿命、繁殖以及生存分布出现了逆转表现为正面刺激作用。

整体上看,无论母代是否经历农药处理,母代低强度的热背景经历,提高了子代种群参数的表现,而母代高强度的热背景经历,虽然刺激了子代的繁殖,但由于延长了发育所需时间,因此对子代种群参数整体仍然产生了负面影响。此外,母代农药处理对子代种群参数的刺激增长作用仅在对照背景下表现出来,而在母代经不同强度的热背景处理后,这种对子代种群参数的刺激作用就消失了。从结果可以看出:尽管母代高强度热背景经历后所产生的负面影响并未延续到子代成蚜表现和生存分布,但对子代整体的种群适合度而言,仍然产生了负面影响。与此同时,母代农药处理后对子代种群参数的负面影响也仅在高低强度的热背景经历下体现出来。

总之,母代经历不同强度的热历史背景经历后均会提高当代存活个体的耐热性,但却对当代生活史性状产生不同的影响,低强度的热背景经历不会引起适合度代价的产生,但高强度的热背景驯化却会产生相当严重的适合度代价,导致当代种群整体表现下降。与此同时,不同热历史背景经历也会影响当代生物对于农药的耐受能力,低强度的热背景经历会增强当代试虫对于农药的耐受性,而高强度的热背景经历使得当代试虫对农药更加敏感,从而加重农药所产生的负面影响。母代经历不同热背景处理不仅会影响试虫当代的生活史性状,而且这种影响甚至会延续到子代,但是延续的影响并不是一成不变的,对当代产生负面影响的处理对子代可能会产生正面刺激作用。且不同热历史背景经历也会影响农药对子代生活史性状的影响,对照背景下,农药的处理不会产生母代效应,但在不同热背景驯化后,农药的影响就会产生跨代效应,显著延长子代发育所需时间。因此,在气候变暖、极端天气增加的大背景下,环境温度历史背景对生物应对其他环境胁迫的影响应受到更多关注。

第 5 章

第 5 章

结　论

1. 亚致死剂量的高温强度、行为反应时间组合产生不同生态学后果

麦长管蚜经历短期 1 d 的三种高温强度、行为反应时间组合 34℃/180 min、36℃/30 min、38℃/10 min 单作处理时,行为反应时间较长的温和和中等高温组合(34℃/180 min 和 36℃/30 min)对当代生活史性状以及子代发育、存活产生的负面影响较大。但随着持续天数的增加,较高高温强度、行为反应时间组合 38℃/10 min 产生的负面影响逐渐增加。

2. 高温强度、行为反应时间组合与吡虫啉同时作用产生复杂的互作效应

温和中等高温、行为反应时间组合与吡虫啉作用产生较显著的加性或协同负面互作效果,如显著增加当代即时死亡率和适合度性状,显著抑制子代若蚜发育历期和存活;但高强度高温、行为反应时间组合和农药互作,对当代以及子代的生活史性状多不

产生显著负面影响,对某些性状甚至会产生有益的拮抗作用,如加速子代发育、提高子代存活率以及刺激子代成蚜繁殖等。

3. 高温、吡虫啉胁迫次序对当代没有显著影响,但产生了显著的跨代效应

麦长管蚜经高温吡虫啉先后不同胁迫次序处理后,会产生不同的互作效应,虽对母代本身后续的生活史影响不大。但会产生不同的跨代效应,具体而言,先农药后高温胁迫次序对子代存活产生的负面影响更加强烈,但对子代其他生活史性状产生的影响较小。而先高温后农药胁迫次序尽管对子代存活产生的负面影响较小,但对子代后续的适合度性状影响较大。

4. 与单一高温胁迫相比,高温、吡虫啉双重胁迫对昆虫影响更显著

与单一高温胁迫相比,无论高温吡虫啉胁迫次序如何,昆虫遭受高温、吡虫啉双重胁迫后均会对生物当代适合度性状产生更严重的负面影响,同时也会产生更加显著的跨代效应,对子代产生显著的影响,既包括负面影响也包括正面刺激作用。

5. 当代高温和吡虫啉胁迫对子代的种群动态产生显著影响

无论是经历单一高温胁迫还是高温、吡虫啉双重胁迫,对子代的负面影响均集中在发育和存活,如延长子代的发育历期,降低子代的若虫存活率,但这种负面影响一般不会延续到子代的成蚜性状,甚至会对子代的成蚜性状产生正面刺激。尽管母代胁迫经历不会影响子代成蚜性状甚至会产生刺激作用,但是由于胁迫对子代发育的负面影响,对子代总体内禀增长率的贡献更大,因此会导致子代种群增长率整体下降。

6. 早期热背景经历影响后期高温和农药响应表现,甚至产生显著跨代效应

当代若蚜期的高、低强度热背景经历,在显著提高成蚜期耐热性的同时,也对成蚜的耐药性产生了显著影响。若蚜期低强度热背景经历提高了成蚜耐药性,而高强度热背景经历使成蚜对农药更加敏感。此外,当代若蚜期热历史背景的影响甚至可以跨越

世代延续到子代。有趣的是,对当代成虫繁殖表现产生负面影响的高强度温度背景经历对子代成虫繁殖表现却有正面的促进作用。且当代热背景经历也会改变吡虫啉对子代生活史性状的影响,当代单独吡虫啉处理不产生跨代效应,但经历热背景驯化后,吡虫啉的影响就会产生跨代效应,如显著延长子代发育所需时间,从而降低种群增长速率。

参考文献

参考文献

[1] FUKAMI T, WARDLE D A. Long-term ecological dynamics: reciprocal insights from natural and anthropogenic gradients [J]. Proceedings Biological Sciences, 2005, 272 (1577): 2105-2115.

[2] FERRARI M C O, MUNDAY P L, Rummer J L, et al. Interactive effects of ocean acidification and rising sea temperatures alter predation rate and predator selectivity in reef fish communities [J]. Global Change Biology, 2015, 21 (5): 1848-1855.

[3] WILLIS K J, BHAGWAT S A. Biodiversity and climate change [J]. Science, 2009, 326 (5954): 806-807.

[4] FISCHER K, KLOCKMANN M, REIM E. Strong negative effects of simulated heat waves in a tropical butterfly [J]. The Journal of Experimental Biology, 2014, 217: 2892-2898.

[5] MA G, RUDOLF V H, Ma C S. Extreme temperature events alter demographic rates, relative fitness, and community structure [J]. Global Change Biology, 2015, 21 (5): 1794-1808.

[6] DEUTSCH C A, Tewksbury J J, Huey R B, et al. Impacts of climate war ming on terrestrial ectotherms across latitude [J]. Proceedings of the National Academy of Sciences 2008, 105 (18):

6668-6672.

[7] BEKETOV M A, KEFFORD B J, SCHÄFER R B, et al. Pesticides reduce regional biodiversity of stream invertebrates [J]. Proceedings of the National Academy of Sciences, 2013, 110(27): 11039-11043.

[8] CLARKE J H, CLARK W S, Hancock M. Strategies for the prevention of development of pesticide resistance in the UK—lessons for and from the use of herbicides, fungicides and insecticides [J]. Pest Management Science, 1997, 51(3): 391-397.

[9] MÜLLER H M, STAN H J. Pesticide residue analysis in food with CGC—study of long—term stability by the use of different injection techniques [J]. Journal of Separation Science, 2015, 13(10): 697-701.

[10] COHEN E. Pesticide-mediated homeostatic modulation in arthropods [J]. Pesticide Biochemistry and Physiology, 2006, 85(1): 21-27.

[11] SCHOWALTER T D. Insect ecology: an ecosystem approach [M]. United States: Academic Press, 2016: 53-59.

[12] HANSEN J, SATO M, RUEDY R. Perception of climate change [J]. Proceedings of the National Academy of Sciences 2012, 109(37): E2415-E2423.

[13] EDENHOFER O. Summary for Policymakers [M]. Edenhofer O, Pichs-Madruga R.: Cambridge University Press, 2013. 120-210.

[14] MEEHL G A, TEBALDI C. More intense, more frequent and longer lasting heat waves in the 21st century [J]. Science, 2004, 305(5686): 994-997.

[15] 郭志梅, 缪启龙, 李雄. 中国北方地区近50年来气温变化特征的研究 [J]. 地理科学, 2005, 25(4): 66-72.

[16] 谭方颖, 王建林, 宋迎波. 华北平原气候变暖对气象灾害发生趋势的影响 [J]. 自然灾害学报, 2010, 19(5): 125-131.

[17] SOUCH C, GRIMMOND C S B. Applied climatology:

"heat waves" [J]. Progress in Physical Geography, 2004, 28（4）: 599-606.

[18] Clusella-Trullas S, Blackburn T M, Chown S L. Climatic predictors of temperature performance curve parameters in ectotherms imply complex responses to climate change [J]. The American Naturalist, 2011, 177（6）: 738-751.

[19] 陈瑜, 马春森. 气候变暖对昆虫影响研究进展 [J]. 生态学报, 2010, 30（8）: 2159-2172.

[20] 马春森, 马罡, 赵飞. 气候变暖对麦蚜的影响 [J]. 应用昆虫学报, 2014, 51（6）: 1435-1443.

[21] Prosser C L, Nelson D O. The role of nervous systems in temperature adaptation of poikilotherms [J]. Annual Review of Physiology, 1981, 43（43）: 281-300.

[22] David J R, Araripe L O, Chakir M, et al. Male sterility at extreme temperatures: a significant but neglected phenomenon for understanding Drosophila climatic adaptations [J]. Journal of Evolutionary Biology, 2010, 18（4）: 838-846.

[23] Karl I, Stoks R, DE B M, et al. Temperature extremes and butterfly fitness: conflicting evidence from life history and immune function [J]. Global Change Biology, 2011, 17（2）: 676-687.

[24] Denlinger D L, Yocum G D. Physiology of heat sensitivity [M]. London: Westview Press Oxford, 1998. 7-53.

[25] Mourier H, Poulsen K P. Control of insects and mites in grain using a high temperature/short time（HTST）technique [J]. Journal of Stored Products Research, 2000, 36（3）: 309-318.

[26] Rasmont P, Iserbyt S. The bumblebees scarcity syndrome: are heat waves leading to local extinctions of bumblebees（Hymenoptera: Apidae: Bombus）? [J]. Annales de la Société entomologique de France, 2012, 48（3-4）: 275-280.

[27] Fields P G. The control of stored-product insects and mites with extreme temperatures [J]. Journal of Stored Products Research, 1992, 28（2）: 89-118.

[28] Kalosaka K, Soumaka E, Politis N, et al. Thermotolerance

and HSP70 expression in the Mediterranean fruit fly *Ceratitis capitata* [J]. Journal of Insect Physiology, 2009, 55（6）: 568-573.

[29] 马骏, 万方浩, 郭建英, 等. 豚草卷蛾对温湿度的适应性 [J]. 中国生物防治, 2003, 19（4）: 158-161.

[30] 王红静, 范惠菊. 高温闷杀法防治温室烟粉虱 [J]. 蔬菜, 2003（10）: 30-31.

[31] 周永丰, 唐峻岭. 高温对南美斑潜蝇的致死作用 [J]. 昆虫知识, 2003, 40（4）: 372-373.

[32] Bahrndorff S, Mariën J, Loeschcke V, et al. Dynamics of heat-induced thermal stress resistance and hsp70 expression in the springtail, *Orchesella cincta* [J]. Functional Ecology, 2009, 23（2）: 233-239.

[33] CHIDAWANYIKA F, TERBLANCHE J S. Rapid thermal responses and thermal tolerance in adult codling moth *Cydia pomonella* (Lepidoptera: Tortricidae) [J]. Journal of Insect Physiology, 2011, 57（1）: 108-117.

[34] York H, Oberhauser K. Effects of duration and timing of heat stress on Monarch Butterfly (*Danaus plexippus*) (Lepidoptera: Nymphalidae) development [J]. Journal of the Kansas Entomological Society, 2002, 75（4）: 290-298.

[35] Asin L, Pons X. Effect of high temperature on the growth and reproduction of corn aphids (Homoptera: Aphididae) and implications for their population dynamics on the northeastern Iberian peninsula [J]. Environment Entomology, 2001, 30（6）: 1127-1134.

[36] Fasolo A G, Krebs R A. A comparison of behavioural change in *Drosophila* during exposure to thermal stress [J]. Biological journal of the Linnean Society, 2004, 83（2）: 197-205.

[37] Hoffmann A A. Physiological climatic limits in *Drosophila*: patterns and implications [J]. The Journal of Experimental Biology, 2010, 213（6）: 870-880.

[38] 曹新民, 邓永学, 赵志模, 等. 温度对四纹豆象生长发

育与繁殖的影响 [J]. 昆虫知识, 2009, 46（2）: 233-237.

[39] PIYAPHONGKUL J, PRITCHARD J, BALE J. Heat stress impedes development and lowers fecundity of the brown planthopper *Nilaparvata lugens*（Stål）[J]. Plos One, 2012, 7(10): e47413.

[40] 马春森, 陈瑞鹿. 温度对小菜蛾（*Plutella xylostella L.*）发育和繁殖影响的研究 [J]. 吉林农业科学, 1993,（3）: 44-49.

[41] ZHANG W, ZHAO F, Hoffmann A A, et al. A single hot event that does not affect survival but decreases reproduction in the diamondback moth, *Plutella xylostella* [J]. Plos One, 2013, 8（10）: e75923.

[42] JØRGENSEN K T, SØRENSEN J G, BUNDGAARD J. Heat tolerance and the effect of mild heat stress on reproductive characters in *Drosophila buzzatii* males [J]. Journal of Thermal Biology, 2006, 31（3）: 280-286.

[43] ROUX O, LE LANN C, VAN ALPHEN J J, et al. How does heat shock affect the life history traits of adults and progeny of the aphid parasitoid *Aphidius avenae*（Hymenoptera: Aphidiidae）? [J]. Bulletin of Entomological Research, 2010, 100（5）: 543-549.

[44] GOMI T, NAGASAKA M, FUKUDA T, et al. Shifting of the life cycle and life-history traits of the fall webworm in relation to climate change. [J]. Entomologia Experimentalis ET Applicata, 2010, 125（2）: 179-184.

[45] MAISTRELLO L, LOMBROSO L, PEDRONI E, et al. Summer raids of Arocatus melanocephalus（Heteroptera, Lygaeidae）in urban buildings in Northern Italy: Is climate change to blame? [J]. Journal of Thermal Biology, 2006, 31（8）: 594-598.

[46] KREBS R A, LOESCHCKE V. Costs and benefits of activation of the heat-shock response in *Drosophila melanogaster* [J]. Functional Ecology, 1994, 8（6）: 730-737.

[47] JEFFS C T, LEATHER S R. Effects of extreme, fluctuating temperature events on life history traits of the grain

aphid, *Sitobion avenae* [J]. Entomologia Experimentalis Et Applicata, 2014, 150（3）: 240-249.

[48] MOUSSEAU T A, DINGLE H. Maternal effects in insect life histories [J]. Annual Review of Entomology, 1991, 36（1）: 511-534.

[49] EZARD T H G, PRIZAK R, Hoyle R B, et al. The fitness costs of adaptation via phenotypic plasticity and maternal effects [J]. Functional Ecology, 2014, 28（3）: 693-701.

[50] FOX C W, CZESAK M E. Evolutionary ecology of progeny size in arthropods [J]. Annual Review of Entomology, 2000, 45（1）: 341-369.

[51] PLAISTOW S J, LAPSLEY C T, Benton T G. Context-dependent intergenerational effects: the interaction between past and present environments and its effect on population dynamics [J]. American Naturalist, 2006, 167（2）: 206-215.

[52] BENTON T G, STCLAIR J J, Plaistow S J. Maternal effects mediated by maternal age: from life histories to population dynamics [J]. Journal of Animal Ecology, 2008, 77（5）: 1038-1046.

[53] MOUSSEAU T A, FOX C W. The adaptive significance of maternal effects [J]. Trends in Ecology and Evolution, 1998, 13（10）: 403-407.

[54] TRIGGS A M, KNELL R J. Parental diet has strong transgenerational effects on offspring immunity [J]. Functional Ecology, 2012, 26（6）: 1409-1417.

[55] ZEHNDER C B, PARRIS M A, Hunter M D. Effects of maternal age and environment on offspring vital rates in the *Oleander aphid*（Hemiptera : Aphididae）[J]. Environmental Entomology, 2007, 36（4）: 910-917.

[56] REINHOLD K. MATERNAL effects and the evolution of behavioral and morphological characters: A literature review indicates the importance of extended maternal care [J]. Journal of Heredity, 2002, 93（6）: 400-405.

[57] HOFFMANN A A, HEWA-KAPUGE S. Acclimation for heat resistance in *Trichogrammanr. brassicae*: can it occur without costs? [J]. Functional Ecology, 2010, 14（1）: 55-60.

[58] BURGESS S C, Marshall D J. Temperature-induced maternal effects and environmental predictability [J]. The Journal of Experimental Biology, 2011, 214（14）: 2329-2336.

[59] ERNSTING G, ISAAKS J. Effects of temperature and season on egg size, hatchling size and adult size in *Notiophilus biguttatus* [J]. Ecological Entomology, 1997, 22（1）: 32-40.

[60] STEIGENGA M, ZWAAN B, Brakefield P, et al. The evolutionary genetics of egg size plasticity in a butterfly [J]. Journal of evolutionary biology, 2005, 18（2）: 281-289.

[61] ROSSITER M. MATERNAL effects generate variation in life history: consequences of egg weight plasticity in the gypsy moth [J]. Functional Ecology, 1991: 386-393.

[62] GILCHRIST G W, HUEY R B. Parental and developmental temperature effects on the thermal dependence of fitness in *Drosophila melanogaster* [J]. Evolution, 2001, 55（1）: 209-214.

[63] ISMAEIL I, DOURY G, Desouhant E, et al. Trans-generational effects of mild heat stress on the life history traits of an aphid parasitoid [J]. Plos One, 2013, 8（2）: e54306.

[64] GUO J Y, CONG L, Zhou Z S, et al. Multi-generation life tables of *Bemisia tabaci*（Gennadius）biotype B（Hemiptera: Aleyrodidae）under high-temperature stress [J]. Environmental entomology, 2012, 41（6）: 1672-1679.

[65] FU X J, MAN S Y, Wen J Y, et al. Influence of continuous high temperature conditions on Wolbachia infection frequency and the fitness of *Liposcelis tricolor*（Psocoptera: Liposcelididae）[J]. Environmental Entomology, 2009, 38（5）: 1365.

[66] 胡战雄, 杨巧英. 理性面对农药在现代农业生产上的应用 [J]. 农业科技与信息, 2018,（20）: 67-70.

[67] DIEKÖTTER T, Crist T O. Quantifying habitat-specific contributions to insect diversity in agricultural mosaic landscapes [J]. Insect Conservation and Diversity, 2013, 6（5）: 607-618.

[68] SONODA S, KOHARA Y, KOSHIYAMA Y, et al. Effects of pesticide practices on insect biodiversity in peach orchards [J]. Applied Entomology and Zoology, 2011, 46（3）: 335-342.

[69] RAUPP M J, HOLMES J J, SADOF C, et al. Effects of cover sprays and residual pesticides on scale insects and natural enemies in urban forests [J]. Journal of Arboriculture, 2001, 27（4）: 203-204.

[70] 史树森, 高月波, 臧连生, 等. 不同杀虫剂对大豆田节肢动物群落结构的影响 [J]. 应用昆虫学报, 2012, 49（5）: 1249-1254.

[71] 刘燕承. 禾谷缢管蚜抗药性及呋虫胺对其亚致死效应研究 [D]. 雅安: 四川农业大学, 2016.

[72] SHI X B, JIANG L L, WANG H Y, et al. Toxicities and sublethal effects of seven neonicotinoid insecticides on survival, growth and reproduction of imidacloprid - resistant cotton aphid, *Aphis gossypii* [J]. Pest Management Science, 2011, 67（12）: 1528.

[73] RIX R R, CUTLER G C. Does multigenerational exposure to hormetic concentrations of imidacloprid precondition aphids for increased insecticide tolerance? [J]. Pest Management Science, 2017, 74（2）: 314-322.

[74] DESNEUX N, DECOURTYE A, Delpuech J. The sublethal effects of pesticides on beneficial arthropods [J]. Annual Review of Entomology, 2007, 52（1）: 81-106.

[75] MOORE A, WARING C P. Sublethal effects of the pesticide diazinon on olfactory function in mature male *Atlantic Salmon Parr* [J]. Journal of Fish Biology, 2010, 48（4）: 758-775.

[76] RAN W, ZHENG H, CHENG Q, et al. Lethal and sublethal effects of a novel cis-nitromethylene neonicotinoid insecticide, cycloxaprid, on *Bemisia tabaci* [J]. Crop Protection,

2016, 83: 15-19.

[77] SETH R K, KAUR J J, RAO D K, et al. Effects of larval exposure to sublethal concentrations of the ecdysteroid agonists RH-5849 and tebufenozide (RH-5992) on male reproductive physiology in *Spodoptera litura* [J]. Journal of Insect Physiology, 2004, 50 (6): 505-517.

[78] 尹飞, 冯夏, 胡珍娣, 等. 氯虫苯甲酰胺对小菜蛾生长发育的亚致死效应研究 [J]. 广东农业科学, 2012, 39 (17): 78-80.

[79] 都振宝. 吡虫啉和噻虫嗪亚致死剂量对荻草谷网蚜生命表参数及取食行为的影响 [D]. 武汉: 华中农业大学, 2012.

[80] WALKER M, STUFKENS M, Wallace A. Indirect non-target effects of insecticides on Tasmanian brown lacewing (*Micromus tasmaniae*) from feeding on lettuce aphid (*Nasonovia ribisnigri*) [J]. Biological Control, 2007, 43 (1): 31-40.

[81] MOSCARDINI V F, GONTIJO P C, MICHAUD J P, et al. Sublethal effects of insecticide seed treatments on two nearctic lady beetles (Coleoptera: Coccinellidae) [J]. Ecotoxicology, 2015, 24 (5): 1152-1161.

[82] MARTINOU A F, SERAPHIDES N, STAVRINIDES M C. Lethal and behavioral effects of pesticides on the insect predator *Macrolophus pygmaeus* [J]. Chemosphere, 2014, 96 (12): 167-173.

[83] BAO H B, LIU S H, GU J H, et al. Sublethal effects of four insecticides on the reproduction and wing formation of brown planthopper, *Nilaparvata lugens* [J]. Pest Management Science, 2010, 65 (2): 170-174.

[84] 孙小玲, 陈威, 杨巧燕, 等. 三种杀虫剂对多异瓢虫的毒力及亚致死效应 [J]. 草地学报, 2016, 24 (5): 1094-1099.

[85] BIONDI A, DESNEUX N, SISCARO G, et al. Using organic-certified rather than synthetic pesticides may not be safer for biological control agents: selectivity and side effects of 14 pesticides on the predator *Orius laevigatus* [J]. Chemosphere,

2012, 87（7）: 803-812.

[86] CHO S R, KOO H N, YOON C, et al. Sublethal effects of flonicamid and thiamethoxam on green peach aphid, *Myzus persicae* and feeding behavior analysis [J]. Journal of the Korean Society for Applied Biological Chemistry, 2011, 54（6）: 889-898.

[87] FARIBA S, PARVIZ S, Moosa S, et al. Lethal and sublethal effects of buprofezin and imidacloprid on the whitefly parasitoid *Encarsia inaron* [J]. Crop Protection, 2012, 32（1358）: 83-89.

[88] 惠婧婧, 刘长仲, 孟银凤, 等. 吡虫啉对豌豆蚜的亚致死效应 [J]. 植物保护, 2009, 35（5）: 86-88.

[89] 宋亮, 章金明, 吕要斌. 茚虫威和高效氯氰菊酯对小菜蛾的亚致死效应 [J]. 昆虫学报, 2013, 56（5）: 521-529.

[90] YIN X H, WU Q J, LI X F, et al. Sublethal effects of spinosad on *Plutella xylostella*（Lepidoptera: Yponomeutidae）[J]. Crop Protection, 2008, 27（10）: 1385-1391.

[91] WEAVER S J, KUMAR A, CHEN M. Recent increases in extreme temperature occurrence over land [J]. Geophysical Research Letters, 2014, 41（13）: 4669-4675.

[92] SANDROCK C, TANADINI L G, PETTIS J S, et al. Sublethal neonicotinoid insecticide exposure reduces solitary bee reproductive success [J]. Agricultural and Forest Entomology, 2014, 16（2）: 119-128.

[93] WANG X Y, YANG Z Q, SHEN Z R, et al. Sublethal effects of selected insecticides on fecundity and wing dimorphism of green peach aphid [J]. Journal of Applied Entomology, 2008, 132（2）: 135-142.

[94] CONWAY H E, KRING T J, MCNEW R. Effect of imidacloprid on wing formation in the cotton aphid（Homoptera: Aphididae）[J]. Florida Entomologist, 2003, 86（4）: 474-476.

[95] MUSSER F R, SHELTON A M. The influence of post-exposure temperature on the toxicity of insecticides to *Ostrinia nubilalis*（Lepidoptera: Crambidae）[J]. Pest Management

Science, 2005, 61（5）: 508-510.

[96] DE BEECK L O, VERHEYEN J, OLSEN K, et al. Negative effects of pesticides under global war ming can be counteracted by a higher degradation rate and thermal adaptation [J]. Journal of Applied Ecology, 2017, 54（6）: 1847-1855.

[97] HOLMSTRUP M, BINDESBØL A M, Oostingh G J, et al. Interactions between effects of environmental chemicals and natural stressors: a review [J]. Science of the Total Environment, 2010, 408（18）: 3746-3762.

[98] NOYES P D, MCELWEE M K, MILLER H D, et al. The toxicology of climate change: environmental conta minants in a war ming world [J]. Environment International, 2009, 35（6）: 971-986.

[99] ARAMBOUROU H, Stoks R. Combined effects of larval exposure to a heat wave and chlorpyrifos in northern and southern populations of the damselfly *Ischnura elegans* [J]. Chemosphere, 2015, 128: 148-154.

[100] HOOPER M J, ANKLEY G T, Cristol D A, et al. Interactions between chemical and climate stressors: A role for mechanistic toxicology in assessing climate change risks [J]. Environmental Toxicology and Chemistry, 2013, 32（1）: 32-48.

[101] MOE S J, DE S K, CLEMENTS W H, et al. Combined and interactive effects of global climate change and toxicants on populations and communities [J]. Environmental Toxicology and Chemistry, 2013, 32（1）: 49-61.

[102] 苏寿, 叶炳辉. 现代医学昆虫学 [M]. 北京: 高等教育出版社, 1996.

[103] MA G, MA C S. EFFECT of acclimation on heat-escape temperatures of two aphid species: implications for estimating behavioral response of insects to climate war ming [J]. Journal of Insect Physiology, 2012, 58（3）: 303-309.

[104] 刘群. 高温对抗性和敏感小菜蛾 AChE 及解毒酶系活性的影响 [D]. 福州: 福建农林大学, 2009.

[105] 张晓燕. 高温对几种菜田昆虫乙酰胆碱酯酶及解毒酶系活性影响的研究 [D]. 福州：福建农林大学, 2010.

[106] JOHNSON D L. Influence of temperature on toxicity of two pyrethroids to grasshoppers（Orthoptera: Acrididae）[J]. Journal of Economic Entomology, 1990, 83（83）: 366–373.

[107] GOEL N K, STOLEN R H, MORGAN S, et al. Glossary of terms for thermal physiology. Second edition. Revised by the commission for thermal physiology of the international union of physiological sciences（IUPS Thermal Commission）[J]. Pflugers Arch, 1987, 410（4-5）: 567–587.

[108] 刘佳, 高占林, 党志红, 等. 不同温度效应杀虫剂诱导对绿盲蝽三种解毒酶活力的影响 [J]. 应用昆虫学报, 2015, 52（3）: 609–615.

[109] 马云华. 几类杀虫剂对麦长管蚜和绿盲蝽的温度系数及温度对主要代谢酶的影响 [D]. 保定：河北农业大学, 2012.

[110] NORMENT B R, CHAMBERS H W. Temperature relationships in organophosphorus poisoning in boll weevils [J]. Journal of Economic Entomology, 1970, 63（2）: 502–504.

[111] CHALFANT R B. CABBAGE Looper: effect of temperature on toxicity of insecticides in the laboratory [J]. Journal of Economic Entomology, 1973, 66（2）: 339–341.

[112] BLUM M S, KEARNS C W. Temperature and the Action of Pyrethrum in the American Cockroach [J]. Journal of Economic Entomology, 1956, 49（6）: 862–865.

[113] SPARKS T C, SHOUR M H, WELLEMEYER E G. Temperature–toxicity relationships of pyrethroids on three lepidopterans [J]. Journal of Economic Entomology, 1982, 75（4）: 643–646.

[114] 顾晓军, 高飞, 田素芬. 短期高温预处理与氟虫腈对小菜蛾3龄幼虫的联合作用 [J]. 中国生态农业学报, 2009, 17（4）: 709–714.

[115] 李贤贤, 马晓丹, 薛明, 等. 不同药剂对韭菜迟眼蕈蚊致毒的温度效应及田间药效 [J]. 北方园艺, 2014,（9）: 125–128.

[116] GE L Q, HUANG L J, YANG G Q, et al. Molecular basis for insecticide-enhanced thermotolerance in the brown planthopper *Nilaparvata lugens* Stål (Hemiptera:Delphacidae) [J]. Molecular Ecology, 2013, 22（22）: 5624-5634.

[117] 游琳琳. 井冈霉素刺激褐飞虱生殖和增强耐热性的分子机制研究 [D]. 扬州 : 扬州大学, 2017.

[118] 苏明明. 高温胁迫对 Q 烟粉虱耐药性的影响及其机理 [D]. 青岛 : 青岛农业大学, 2016.

[119] RAGHAVENDRA K, BARIK T K, ADAK T. Development of larval thermotolerance and its impact on adult susceptibility to malathion insecticide and Plasmodium vivax infection in *Anopheles stephensi* [J]. Parasitology Research, 2010, 107（6）: 1291-1297.

[120] PATIL N S, LOLE K S, DEOBAGKAR D N. Adaptive larval thermotolerance and induced cross - tolerance to propoxur insecticide in mosquitoes *Anopheles stephensi* and *Aedes aegypti* [J]. Medical and Veterinary Entomology, 1996, 10（3）: 277-282.

[121] 王德辉. 短期高温与农药对小菜蛾幼虫相互作用的研究 [D]. 福州 : 福建农林大学, 2008.

[122] FENG H, WANG L, LIU Y, et al. Molecular characterization and expression of a heat shock protein gene (Hsp 90) from the car mine spider mite, *Tetranychus cinnabarinus* (Boisduval) [J]. Journal of Insect Science, 2010, 10（112）: 112.

[123] SONODA S, TSUMUKI H. Induction of heat shock protein genes by chlorfenapyr in cultured cells of the cabbage armyworm, *Mamestra brassicae* [J]. Pesticide Biochemistry and Physiology, 2007, 89（3）: 185-189.

[124] SUN Y, SHENG Y, BAI L, et al. Characterizing heat shock protein 90 gene of *Apolygus lucorum* (Meyer-Dür) and its expression in response to different temperature and pesticide stresses [J]. Cell Stress and Chaperones, 2014, 19（5）: 725-739.

[125] CHEN J, KITAZUMI A, ALPUERTO J, et al. Heat-induced mortality and expression of heat shock proteins in

Colorado potato beetles treated with imidacloprid [J]. Insect Science, 2016, 23（4）: 548-554.

[126] MUTURI E J, LAMPMAN R L, COSTANZO K, et al. Effect of temperature and insecticide stress on life-history traits of *Culex restuans* and *Aedes albopictus*（Diptera: Culicidae）[J]. Journal of Medical Entomology, 2011, 48（2）: 243-250.

[127] WANG X Y, SHEN Z R. Potency of some novel insecticides at various environmental temperatures on *Myzus persicae* [J]. Phytoparasitica, 2007, 35（4）: 414-422.

[128] JANSSENS L, STOKS R. Chlorpyrifos-induced oxidative damage is reduced under war ming and predation risk: explaining antagonistic interactions with a pesticide [J]. Enviro Pollut, 2017, 226（Supplement C）: 79-88.

[129] CHENG J, HUANG L J, ZHU Z F, et al. Heat-dependent fecundity enhancement observed in *Nilaparvata lugens*（Hemiptera: Delphacidae）after treatment with triazophos [J]. Environmental Entomology, 2014, 43（2）: 474-481.

[130] YU Y, HUANG L, WANG L, et al. The combined effects of temperature and insecticide on the fecundity of adult males and adult females of the brown planthopper *Nilaparvata lugens* Stål（Hemiptera: Delphacidae）[J]. Crop Protection, 2012, 34: 59-64.

[131] JANSSENS L, TÜZÜN N, STOKS R. Testing the time-scale dependence of delayed interactions: A heat wave during the egg stage shapes how a pesticide interacts with a successive heat wave in the larval stage[J]. Environmental Pollution, 2017, 230（Supplement C）: 351-359.

[132] JANSSENS L, DINH V K, STOKS R. Extreme temperatures in the adult stage shape delayed effects of larval pesticide stress: a comparison between latitudes [J]. Aquatic Toxicology, 2014, 148（2）: 74-82.

[133] LI H, ZHENG Y, SUN D, et al. Combined effects of temperature and avermectins on life history and stress response of the western flower thrips, *Frankliniella occidentalis*[J]. Pesticide

Biochemistry and Physiology, 2014, 108: 42-48.

[134] 左太强, 张彬, 张绍婷, 等. 高温和啶虫脒处理西花蓟马对其 F1 代生命表参数的联合作用[J]. 昆虫学报, 2015, (7): 767-775.

[135] FOLGUERA G, BASTÍAS D A, CAERS J, et al. An experimental test of the role of environmental temperature variability on ectotherm molecular, physiological and life-history traits: Implications for global war ming [J]. Comparative Biochemistry and Physiology-Part A: Molecular & Integrative Physiology, 2011, 159 (3): 242-246.

[136] XING K, HOFFMANN A A, MA C S. Does thermal variability experienced at the egg stage influence life history traits across life cycle stages in a small invertebrate? [J]. PLoS One, 2014, 9 (6): e99500.

[137] XING K, MA C S, ZHAO F, et al. Effects of large temperature fluctuations on hatching and subsequent development of the diamondback moth (Lepidoptera: Plutellidae) [J]. Florida Entomologist, 2015, 98 (2): 651-659.

[138] ZHAO F, HOFFMANN A A, XING K, et al. Life stages of an aphid living under similar thermal conditions differ in thermal performance [J]. Journal of Insect Physiology, 2017, 99: 1-7.

[139] CUI X, WAN F, XIE M, et al. Effects of heat shock on survival and reproduction of two whitefly species, *Trialeurodes vaporariorum* and *Bemisia tabaci* biotype B [J]. Journal of Insect Science, 2008, 8 (24): 1-10.

[140] PIETERSE W, TERBLANCHE J S, ADDISON P. Do thermal tolerances and rapid thermal responses contribute to the invasion potential of *Bactrocera dorsalis* (Diptera: Tephritidae)? [J]. Journal of Insect Physiology, 2017, 98: 1-6.

[141] KEARNEY M, SHINE R, PORTER W P. The potential for behavioral thermoregulation to buffer "cold-blooded" animals against climate war ming [J]. Proceedings of the National Academy

of Sciences, 2009, 106（10）: 3835.

[142] PINCEBOURDE S, SINOQUET H, CASAS J. Regional climate modulates the canopy mosaic of favourable and risky microclimates for insects [J]. Journal of Animal Ecology, 2007, 76（3）: 424-438.

[143] MA G, MA C S. Climate war ming may increase aphids' dropping probabilities in response to high temperatures [J]. Journal of Insect Physiology, 2012, 58（11）: 1456-1462.

[144] ANDERSEN D H, PERTOLDI C, Scali V, et al. Heat stress and age induced maternal effects on wing size and shape in parthenogenetic *Drosophila mercatorum* [J]. Journal of Evolutionary Biology, 2005, 18（4）: 884-892.

[145] KINGSOLVER J, BUCKLEY L. Evolution of plasticity and adaptive responses to climate change along climate gradients [J]. Proceedings of the National Academy of Sciences, 2017, 284（1860）:386.

[146] 白莉, 郑王义, 任东植, 等. 麦长管蚜为害损失估计及防治阈值研究 [J]. 山西农业科学, 2006, 34（1）: 61-64.

[147] BROWN J H, GILLOOLY J F, ALLEN A P, et al. Toward a metabolic theory of ecology [J]. Ecology, 2004, 85（7）: 1771-1789.

[148] ANGILLETTA M J. Thermal adaptation: a theoretical and empirical synthesis [M]. New York: Oxford University Press, 2009. 35-53.

[149] DIXON A F G. Aphid ecology: an optimization approach [M]. Germany: Springer-Verlag, 1998. 59-81.

[150] DANKS H V. Short life cycles in insects and mites [J]. Canadian Entomologist, 2006, 138（4）: 407-463.

[151] 曹雅忠, 尹姣, 李克斌, 等. 小麦蚜虫不断猖獗原因及控制对策的探讨 [J]. 植物保护, 2006, 32（05）: 72-75.

[152] NAUEN R, HUNGENBERG H, TOLLO B, et al. Antifeedant effect, biological efficacy and high affinity binding of imidacloprid to acetylcholine receptors in *Myzus persicae* and

Myzus nicotianae [J]. Pest Management Science, 2015a, 53（2）: 133–140.

[153] NAUEN R, TIETJEN K, WAGNER K, et al. Efficacy of plant metabolites of imidacloprid against *Myzus persicae* and *Aphis gossypii*（Homoptera: Aphididae）[J]. Pest Management Science, 2015b, 52（1）: 53–57.

[154] SENEVIRATNE S I, DONAT M G, Mueller B, et al. No pause in the increase of hot temperature extremes [J]. Nature Climate Change, 2014, 4（3）: 161–163.

[155] WANG S Y S, YOON J H, FUNK C C, et al. Recent increases in extreme temperature occurrence over land[M].John Wiley & Sons, Inc., 2017.

[156] DILLON M E, CAHN L R Y, HUEY R B. Life history consequences of temperature transients in *Drosophila melanogaster* [J]. Journal of Experimental Biology, 2007, 210（16）: 2897–2904.

[157] DEBARRO P J, MAELZER D A. Influence of high-temperatures on the survival of *Rhopalosiphum Padi*（L）（Hemiptera, Aphididae）in irrigated perennial grass pastures in South-Australia [J]. Australian Journal of Zoology, 1993, 41（2）: 123–132.

[158] CHAPPERON C, SEURONT L. Behavioral thermoregulation in a tropical gastropod: links to climate change scenarios [J]. Global Change Biology, 2011, 17（4）: 1740–1749.

[159] WIKTELIUS S. DISTRIBUTION of *Rhopalosiphum padi*（Homoptera: Aphididae）on spring barley plants [J]. Annals of Applied Biology, 2010, 110（1）: 1–7.

[160] 韩巨才,刘慧平,徐建岗,等.山西麦长管蚜对拟除虫菊酯杀虫剂抗药性研究 [J].山西农业科学,1996,（2）: 26–28.

[161] 魏岑,黄绍宁,范贤林,等.麦长管蚜的抗药性研究 [J].昆虫学报,1988, 31（2）: 148–156.

[162] ZHAO F, ZHANG W, HOFFMANN A A, et al. Night warming on hot days produces novel impacts on development, survival

and reproduction in a small arthropod [J]. Journal of Animal Ecology, 2014, 83（4）: 769-778.

[163] MANSOUR N A, ELDEFRAWI M E, TOPPOZADA A, et al. Toxicological studies on the Egyptian cotton leaf worm, *Prodenia litura*. VI. Potentiation and antagonism of organophosphorus and carbamate Insecticides [J]. Journal of Economic Entomology, 1966, 59（2）: 307-311.

[164] GU X J, TIAN S F, WANG D H, et al. Interaction between short-term heat pretreatment and fipronil on 2 Instar Larvae of Diamondback Moth, *Plutella Xylostella*（Linn）[J]. Dose-response 2010, 8（3）: 331.

[165] 张伟, 吕利华, 何余容, 等. 玫烟色棒束孢与短期35℃高温对小菜蛾3龄幼虫的联合作用 [J]. 环境昆虫学报, 2013, 35（02）: 182-189.

[166] STOKS R, BLOCK M D, McPeek M A. Physiological costs of compensatory growth in a damselfly [J]. Ecology, 2006, 87（6）: 1566-1574.

[167] DMITRIEW C, ROWE L. EFFECTS of early resource limitation and compensatory growth on lifetime fitness in the ladybird beetle（*Harmonia axyridis*）[J]. J Evolution Biol, 2007, 20（4）: 1298-1310.

[168] CAMPERO M, DE B M, Ollevier F, et al. Metamorphosis offsets the link between larval stress, adult asymmetry and individual quality [J]. Functional Ecology, 2008, 22（2）: 271-277.

[169] POTTER K A, DAVIDOWITZ G, Arthur W H. Cross-stage consequences of egg temperature in the insect *Manduca sexta* [J]. Functional Ecology, 2011, 25（3）: 548-556.

[170] LIANG L N, ZHANG W, Ma G, et al. A single hot event stimulates adult performance but reduces egg survival in the Oriental Fruit Moth, *Grapholitha molesta* [J]. Plos One, 2014, 9（12）: e116339.

[171] WANG S Y, QI Y F, DESNEUX N, et al. Sublethal and transgenerational effects of short-term and chronic exposures to the neonicotinoid nitenpyram on the cotton aphid *Aphis gossypii* [J]. Journal of Pest Science, 2017, 90（1）: 389-396.

[172] 杜尧, 马春森, 赵清华, 等. 高温对昆虫影响的生理生化作用机理研究进展 [J]. 生态学报, 2007, 27（4）: 1565-1572.

[173] PRANGE H D. Evaporative cooling in insects [J]. Journal of Insect Physiology, 1996, 42（5）: 493-499.

[174] COHEN A C, Patana R. Ontogenetic and stress-related changes in hemolymph chemistry of beet armyworms [J]. Comparative Biochemistry and Physiology Part A Physiology, 1982, 71（2）: 193-198.

[175] HARGIS M T, STORCK C W, WICKSTROM E, et al. Hsp27 anti-sense oligonucleotides sensitize the microtubular cytoskeleton of Chinese hamster ovary cells grown at low pH to 42 degrees C-induced reorganization [J]. Int J Hyperthermia, 2004, 20（5）: 491-502.

[176] RENSING L, RUOFF P. Temperature effect on entrainment, phase shifting, and amplitude of circadian clocks and its molecular bases [J]. Chronobiology International, 2002, 19（5）: 807.

[177] STANLEY K, FENTON B. A member of the Hsp60 gene family from the peach potato aphid, *Myzus persicae*（Sulzer.）[J]. Insect Molecular Biology, 2000, 9（2）: 211-215.

[178] 李继. 有机磷杀虫剂对乙酰胆碱 M2 受体磷酸化抑制作用的研究 [D]. 长春: 吉林大学, 2007.

[179] 王园园. 作用于钠通道的杀虫剂对 DSC1 通道的影响及透骨草作用机理初探 [D]. 咸阳: 西北农林科技大学, 2016.

[180] 刘佳. 杀虫剂对绿盲蝽毒力的温度效应及其解毒酶相关机制研究 [D]. 保定: 河北农业大学, 2015.

[181] JACOBSON T, PREVODNIK A, SUNDELIN B. Combined effects of temperature and a pesticide on the Baltic amphipod *Monoporeia affinis* [J]. Aquatic Biology, 2008, 1（3）:

269-276.

[182] OSTERAUER R, KÖHLER H R. Temperature-dependent effects of the pesticides thiacloprid and diazinon on the embryonic development of zebrafish (*Danio rerio*) [J]. Aquatic Toxicology, 2008, 86 (4): 485-494.

[183] ASHAUER R, HINTERMEISTER A, CARAVATTI I, et al. Toxicokinetic and toxicodynamic modeling explains carry-over toxicity from exposure to diazinon by slow organism recovery [J]. Environmental Science & Technology, 2010, 44 (10): 3963-3971.

[184] Gibbs M, Breuker C, HESKETH H, et al. Maternal effects, flight versus fecundity trade-offs, and offspring immune defence in the Speckled Wood butterfly, *Pararge aegeria* [J]. BMC evolutionary biology, 2010, 10 (1): 345.

[185] GILBERT S F. Mechanisms for the environmental regulation of gene expression: Ecological aspects of animal development [J]. Journal of Biosciences, 2005, 30 (1): 65.

[186] MONAGHAN P. Early growth conditions, phenotypic development and environmental change [J]. Philosophical Transactions of the Royal Society B: Biological Sciences, 2008, 363 (1497): 1635-1645.

[187] LINDSTRÖM J. Early development and fitness in birds and mammals [J]. Trends in Ecology and Evolution, 1999, 14 (9): 343.

[188] METCALFE N B, MONAGHAN P. Compensation for a bad start: grow now, pay later? [J]. Trends in Ecology & Evolution, 2001, 16 (5): 254-260.

[189] HARE K M, LONGSON C G, PLEDGER S, et al. Size, growth, and survival are reduced at cool incubation temperatures in the temperate Lizard *Oligosoma suteri* (Lacertilia: Scincidae) [J]. Copeia, 2004 (2): 383-390.

[190] KAPLAN R H, PHILLIPS P C. Ecological and

developmental context of natural selection: maternal effects and thermally induced plasticity in the frog *Bombina orientalis* [J]. Evolution, 2006, 60（1）: 142-156.

[191] KALRA B, TAMANG A M, PARKASH R. Cross-tolerance effects due to adult heat hardening, desiccation and starvation acclimation of tropical drosophilid-Zaprionus indianus[J]. Comparative Biochemistry and Physiology Part A Molecular & Integrative Physiology, 2017,4（14）: 65-73.

[192] TODGHAM A E, SCHULTE P M, IWAMA G K. Cross-tolerance in the tidepool sculpin: the role of heat shock proteins [J]. Physiological and Biochemical Zoology, 2005, 78（2）: 133-144.

[193] NYAMUKONDIWA C, Terblanche J S. Within-generation variation of critical thermal limits in adult Mediterranean and Natal fruit flies Ceratitis capitata and Ceratitis rosa: thermal history affects short-term responses to temperature [J]. Physiological Entomology, 2010, 35（3）: 255-264.

[194] SCHEIL A E, KÖHLER H, TRIEBSKORN R. Heat tolerance and recovery in Mediterranean land snails after pre-exposure in the field [J]. Journal of Molluscan Studies, 2011, 77（2）: 165-174.

[195] LITTLEWOOD D T J. Thermal tolerance and the effects of temperature on air-gaping in the mangrove oyster, Crassostrea rhizophorae [J]. Comparative Biochemistry and Physiology Part A: Physiology, 1989, 93（2）: 395-397.

[196] KOSH R J, HUTCHISON V H. Thermal tolerances of parietalectomized Anolis carolinensis acclimatized at different temperatures and photoperiods [J]. Herpetologica, 1972, 28（3）: 183-191.

[197] MCKECHNIE A, Chetty K, Lovegrove B. Phenotypic flexibility in the basal metabolic rate of laughing doves: responses to short-term thermal acclimation [J]. The Journal of Experimental Biology, 2007, 210（Pt 1）: 97-106.

[198] WANG L, YANG S, HAN L, et al. Expression profile of two HSP70 chaperone proteins in response to extreme thermal acclimation in Xestia c-nigrum (Lepidoptera: Noctuidae) [J]. Florida Entomologist, 2015, 98 (2): 506-515.

[199] KREBS R A, FEDER M E. Experimental manipulation of the cost of thermal acclimation in Drosophila melanogaster [J]. Biological Journal of The Linnean Society, 1998, 63 (4): 593-601.

[200] MINOIS N. Longevity and aging: beneficial effects of exposure to mild stress [J]. Biogerontology, 2000, 1 (1): 15-29.

[201] SCANNAPIECO A C, SØRENSEN J G, LOESCHCKE V, et al. Heat-induced hormesis in longevity of two sibling Drosophila species [J]. Biogerontology, 2007, 8 (3): 315-325.

[202] PIYAPHONGKUL J, PRITCHARD J, BALE J. Effects of acclimation on the thermal tolerance of the brown planthopper Nilaparvata lugens (Stål) [J]. Agricultural and Forest Entomology, 2014, 16 (2): 174-183.

[203] HOFFMANN A A, SØRENSEN J G, LOESCHCKE V. Adaptation of Drosophila to temperature extremes: bringing together quantitative and molecular approaches [J]. Journal of Thermal Biology, 2003, 28 (3): 175-216.

[204] GIBBS M, BREUKER C J, VAN D. Flight during oviposition reduces maternal egg provisioning and influences offspring development in Pararge aegeria (L.) [J]. Physiological Entomology, 2010, 35 (1): 29-39.